Topology and the Language of Mathematics

Chris Cunliffe

ISBN: 978-0-615-22102-1

©2008, Chris Cunliffe

Bobo Strategy contact information:
Email: info@bobostrategy.com
Website: www.bobostrategy.com

Mailing address:
Bobo Strategy
2506 N Clark # 287
Chicago, IL 60614

All rights reserved. No part of this book may be reproduced without permission in writing from the publisher, Bobo Strategy.

Library of Congress Control Number: 2008906150

Printed in the U.S.A. (Charleston, SC)

Contents

I Preliminary Material   11

1 Logic   13

2 Sets   17

3 Functions   27

4 Exam   39

II Definition of Topology   41

1 Topology   43

2 Standard Topology on $\mathbb{R}$ 49

3 Cofinite Topology   53

4 Closed Sets   59

5 Some Useful Tools   65

6 Some Notes on the Definition of a Topology   69

III Homeomorphisms   71

1 Basis for a Topology   73

2 Continuous Functions   79

3 Homeomorphisms   87

IV More Properties   93

1 Retractions   95

2 Fixed Point Property   99

3 Connectedness   103

V More Examples   111

1 Subspace Topology   113

2 Lower Limit Topology, K-Topology   119

3 Sierpinski Space   123

4 Path Connected   127

VI Separation axioms   131

1 T_1 133

2 T_2 - Hausdorff Spaces   137

3 T_3 - Regular Spaces   143

4 Compact Spaces   145

5 Metric Spaces   149

Acknowledgements

Thank you to the people who helped make both me and this book more useful. I've been lucky to have had some excellent teachers, both in and out of the classroom, and this book is written in appreciation of them. I'm better at everything I do because of them.

Thanks.

Introduction

This book introduces the language of mathematics through point-set topology. Little background in mathematics is assumed.

It is a useful addition to current literature because:

1) The introduction of point-set topology for a primary audience with little to no background in the subject is more effective than some relevant literature in wide use today.

2) The introduction to the language of mathematics is more accessible to the undergraduate / advanced high school math student than some relevant literature in wide use today.

3) It serves as an excellent accompanying text to existing relevant literature.

I hope you find it useful.

- Chris

Part I

Preliminary Material

Chapter 1

Logic

1.1 Remark

X is a car implies X is a car or truck. That is, X is a car $\implies$ X is a car or truck. But X is a car or truck does not imply that X is a car. For example, X could be a truck.

1.2 Remark

X is a car and X is blue $\implies$ X is blue. But X is blue does not imply that X is a car and X is blue. For example, X could be a blue truck.

1.3 Remark

Suppose that Bob and Mary have had only one child and that child's name is Lucy. Then X is Bob and Mary's child $\implies$ X is named Lucy. But X is named Lucy does not imply that X is Bob and Mary's child.

1.4 Remark

X is the state capital of Illinois implies that X is Springfield, IL.
X is Springfield, IL implies that X is the state capital of Illinois.
We can say this more concisely in the following way:
X is the state capital of Illinois if and only if X is Springfield.
X is the state capital of Illinois $\iff$ X is Springfield.

If we want to show that A is true if and only if B is true, we will often show that A implies B and that B implies A.

1.5 Remark

The negation of the statement 'Cindy is a cat' is 'Cindy is not a cat.' The negation of 'There exists an x in T such that x is blarg.' is 'There does not exist an x in T such that x is blarg.' or equivalently, 'Every x in T is not blarg.'

1.6 Remark

Suppose we want to show that every person in the world has the quality of being blarg. Then it is enough to choose one arbitrary person in the world, and show that person must have the quality of being blarg.

For example: Suppose we want to show that every integer is a rational number. Let x be an integer. Then $x = \frac{x}{1}$. $\frac{x}{1}$ is rational. Given an arbitrary integer x, we have shown that x must be rational. So every integer is rational.

1.7 Remark

Let P be the following statement: A implies B. The contrapositive of P is the following statement: not B implies not A. A statement is true if and only if its contrapositive is true. Sometimes if you want to show that A implies B, it is easier to show that not B implies not A. For example: The contrapositive of 'John is blarg implies Jason is bloog' is 'Jason is not bloog implies John is not blarg.'

1.8 Remark

Let x be a rational number and let y be an irrational number. Show $x + y$ is an irrational number.

Proof. Since x is rational, $x = \frac{a}{b}$ for some integers a and b, $b \neq 0$. Suppose that $x + y$ is rational. Then $x + y = \frac{c}{d}$ for some integers c and d, $d \neq 0$. So $\frac{a}{b} + y = \frac{c}{d}$. So $y = \frac{c}{d} - \frac{a}{b}$. So $y = \frac{bc-ad}{bd}$. $a, b, c,$ and d are integers. So $bc, ad, bd,$ and $bc - ad$ are integers. So, we have shown that y is rational. But y is not rational. So our assumption that $x + y$ is rational must have been false. So $x + y$ is irrational. □

This is a proof by contradiction. We assume that something is false and arrive at a contradiction. Then we conclude that what we assumed is false is actually true. We will try not to use proof by contradiction very often, since it is a little awkward.

Chapter 2

Sets

2.1 Definition

A **set** X is a collection of things.

If x is one of the things in X, then x is said to be an **element** of X. This is written $x \in X$. If x is not an element of X we write $x \notin X$. If x and y are elements of X, we write $x \in X$ and $y \in X$ or equivalently $x, y \in X$.

2.2 Example

The set of people whose first name starts with T is a set. We might write this set as: $A = \{x;\ x$ is a person with a name that starts with the letter T$\}$ We read this as 'A equals the set of all x such that x is a person with a name and this name starts with the letter T.' (Notice the ; is read 'such that')

My first name is Chris, so I am not an element of this set. So Chris $\notin A$. Suppose you choose a person named Tom. Then he is in this set. We can say Tom $\in A$. Note that A contains millions of elements.

2.3 Example

Let $X = \{0, 1, 2\}$. This set has three elements. Its three elements are the numbers 0, 1, and 2. $0, 1, 2 \in X$. $3 \notin X$, car $\notin X$, and $2.012 \notin X$.

2.4 Definition

Suppose we have two sets, A and B. Suppose every element of A is an element of B. Then we say that A is a **subset** of B (or A is **contained** in B). We write $A \subset B$.

2.5 Notation

For the rest of the book, let $\mathbb{R}$ = the set of all real numbers, $\mathbb{Z}$ = the set of all integers, $\mathbb{N}$ = the set of all positive integers, $\mathbb{Q}$ = the set of all rational numbers.

2.6 Example

$\mathbb{N} \subset \mathbb{Z} \subset \mathbb{Q} \subset \mathbb{R}$

2.7 Notation

For the rest of this book $\exists$ means 'there exists' and $\forall$ means 'for all'.

2.8 Definition

Let A be the set with no elements. Then A is called the **empty set**. The empty set is often denoted ϕ.

2.9 Example

Let $A = \{x \in \mathbb{R};\ x+2 > -2 \text{ and } x+3 < -5\}$

Claim: $A = \phi$

Proof. Assume that $A \neq \phi$. Then $\exists\, t \in A$.
So $t+2 > -2$ and $t+3 < -5$.
So $t > -4$ and $t < -8$.
This is bad since there isn't any such $t \in \mathbb{R}$.
So our assumption that $A \neq \phi$ has led to a contradiction.
So $A = \phi$.

□

2.10 Result

Let X be any set.

Claim: $\phi \subset X$

Proof. Suppose that ϕ is not contained in X. That means $\exists\, x \in \phi$ such that $x \in X$. So $\exists\, x \in \phi$. But there is no $x \in \phi$. So our assumption that ϕ is not contained in X is false. So $\phi \subset X$.

□

2.11 Example

Let $X = \{0, 1\}$. Then $\phi \subset X$ (by 2.10). But $\phi \notin X$. X has only two elements, the numbers 0 and 1. ϕ (along with every other set) is not an element of X.

2.12 Example

Let $X = \{\phi, 0\}$. Then $\phi \subset X$ by 2.10. X has two elements, the number 0 and the set ϕ. So $\phi \in X$ and $0 \in X$.

2.13 Remark

This empty set stuff is a little awkward. We will use it because it will be useful to us. It helps in set theory similarly to how having the number 0 and negative numbers in arithmetic is sometimes useful.

2.14 Definition

Let A and B be sets.

Then $A \bigcup B = \{x;\ x \in A \text{ or } x \in B\}$.
This is referred to as the **union** of A and B, or A union B.

Then $A \bigcap B = \{x;\ x \in A \text{ and } x \in B\}$.
This is referred to as the **intersection** of A and B, or A intersect B.

2.15 Example

$\mathbb{Q} \bigcap \mathbb{Z} = \mathbb{Z}$, $\mathbb{Q} \bigcup \mathbb{Z} = \mathbb{Q}$, $\mathbb{Z} \bigcap \mathbb{N} = \mathbb{N}$, $\mathbb{Z} \bigcup \mathbb{N} = \mathbb{Z}$.

2.16 Example

If $A = \{1, 2, 3, 5, 8, 13, 21, \ldots\}$ and $B = \{0, 1, 2, 3, 4, 5\}$ then $A \bigcap B = \{1, 2, 3, 5\}$ and $A \bigcup B = \{0, 1, 2, 3, 4, 5, 8, 13, 21, \ldots\}$

2.17 Remark

In 2.16 we did not list every element of A. A had an infinite number of elements, and we described them in a way we hoped was clear. Another possible way to describe the same set A would have been the following. Let $n_1 = 1$. Let $n_2 = 2$. For each integer $i \geq 3$, let $n_i = n_{i-1} + n_{i-2}$. Then define $A = \{n_k;\ k \in \mathbb{N}\}$. Whenever you must describe a set, you should choose a way to correctly describe the elements in a set, concisely and clearly. Sometimes it may be logical to list every element of a set. Sometimes it may make more sense to not.

2.18 Definition

Let A and B be sets. $A = B$ means $A \subset B$ and $B \subset A$.

2.19 Example

Let $A = \{\text{cat, dog}\}$, $B = \{\text{parrot, dog}\}$.
$A \cap B = \{\text{dog}\}$
$A \cup B = \{\text{cat, dog, parrot}\}$

Claim: $\{\text{cat, dog, parrot}\} = \{\text{cat, dog, parrot, cat}\}$

Proof. Let the set on the left hand side of the equals sign be LHS. Let the set on the right hand side of the equals sign be RHS.

LHS $\subset$ RHS:
Every element of LHS is an element of RHS. So LHS $\subset$ RHS.
RHS $\subset$ LHS:
Every element of RHS is an element of LHS. So RHS $\subset$ LHS.

We have shown that LHS $\subset$ RHS and RHS $\subset$ LHS. So LHS = RHS.

$\square$

2.20 Definition

Let X and A be sets. Define $X - A = \{t; t \in X \text{ and } t \notin A\}$. $X - A$ is called the **complement** of A in X.

2.21 Example

$\{0, 1\} \subset \mathbb{R}$

$\mathbb{R} - \{0, 1\} = (-\infty, 0) \bigcup (0, 1) \bigcup (1, \infty)$, where
$(-\infty, 0) = \{x \in \mathbb{R}; x < 0\}$,
$(0, 1) = \{x \in \mathbb{R}; 0 < x < 1\}$ and
$(1, \infty) = \{x \in \mathbb{R}; x > 1\}$.

2.22 Example

Let $A = \{x \in \mathbb{R}; 0 \leq x < 1\}$, $B = \{x \in \mathbb{R}, \frac{1}{2} \leq x \leq 3\}$

We can also write this $A = [0, 1), B = [\frac{1}{2}, 3]$.

Claim: $A \bigcap B = [\frac{1}{2}, 1)$.

Proof. LHS $\subset$ RHS:
Let $x \in A \bigcap B$.
Then $x \in A$ and $X \in B$.
So $0 \leq x < 1$ and $\frac{1}{2} \leq x \leq 3$.
So $x \geq 0, x \geq \frac{1}{2}, x < 1, x \leq 3$.
So $x \geq \frac{1}{2}$ and $x < 1$. So $x \in [\frac{1}{2}, 1)$.

RHS $\subset$ LHS:
Let $x \in [\frac{1}{2}, 1)$.
Then $x \geq 0$ and $x < 1$. So $x \in A$.
And since $x \geq \frac{1}{2}$ and $x \leq 3, x \in B$.
So $x \in A \bigcap B$.
(Why are we done?)

$\square$

2.23 Example

Let $B = \mathbb{R} - \mathbb{Q}$ (the set of irrational numbers).
$\mathbb{Q} \bigcup B = \mathbb{R}$ and $\mathbb{Q} \bigcap B = \phi$

2.24 Problem

Let $X = \mathbb{N}, A = \{2, 4, 6, 8, \ldots\}$
What is $X - A$? Describe this set in three different ways.

2.25 Problem

Let A and B be defined as in 2.22. Show $A \bigcup B = [0, 3]$.

2.26 Definition

If X has exactly n elements for some $n \in \mathbb{N} \bigcup \{0\}$, then we say X is **finite** .

2.27 Example

$\{0, 1, 2, 10\}$ is finite (with 4 elements).
$\mathbb{R}$ and $\mathbb{N}$ are not finite (Why?).
ϕ is finite (with 0 elements).

2.28 Definition

If a set X is not finite, then we say that X is **infinite** .

2.29 Example

$\mathbb{R}$ and $\mathbb{N}$ are infinite.

2.30 Result

Claim: The complement of a union of sets is the intersection of the complements of those sets. (DeMorgan)

Proof. LHS $\subset$ RHS: Let x be in the complement of a union of sets. Then x is not in the union of these sets. So x is not in any of the sets. So, for every set, x is in its complement. So x is in the intersection of the complements of all the sets.

RHS $\subset$ LHS:
Let x be in the intersection of the complements of the sets. Then x is in the complement of each of the sets. So, x is not in any of the sets. So x is not in the union of the sets. So x is in the complement of the union of the sets.

$\square$

2.31 Problem

Show that the complement of an intersection of sets is the union of the complements of those sets. (DeMorgan)

2.32 Result

Let $A \subset X$.

Claim: $X - (X - A) = A$.

Proof. Let $x \in X - (X - A)$. Then $x \in X$ and $x \notin X - A$, so $x \in A$. Let $x \in A$. Then $x \notin X - A$. Since $A \subset X, x \in X$. So $x \in X - (X - A)$. (Why are we done?)

$\square$

2.33 Problem

Let A, B, and C be sets.
Show $(A \cap B) \cup C = A \cap (B \cup C) \iff C \subset A$.

Proof. $\implies$
Suppose $(A \cap B) \cup C = A \cap (B \cup C)$. We want to show that $C \subset A$. So, let $x \in C$. We want to show that $x \in A$. Since $x \in C$, $x \in (A \cap B) \cup C$. So $x \in A \cap (B \cup C)$. So $x \in A$.

$\impliedby$ Suppose $C \subset A$.
We want to show $(A \cap B) \cup C = A \cap (B \cup C)$.

LHS $\subset$ RHS:
Let $x \in (A \cap B) \cup C$. Then $x \in (A \cap B)$ or $x \in C$.

Case 1: $x \in A \cap B$
Then $x \in A$ and $x \in B$. Since $x \in B, x \in B \cup C$. So $x \in A \cap (B \cup C)$.

Case 2: $x \in C$.
Then $x \in B \cup C$. Since $C \subset A, x \in A$. So $x \in A \cap (B \cup C)$.
So we have shown that when we suppose $C \subset A$,

$(A \cap B) \cup C \subset A \cap (B \cup C)$.

There is one thing left to show ... notice what that is, and then show it.

□

Chapter 3

Functions

3.1 Definition

Let A and B be sets. A **function** f from A to B is a rule that assigns to every $a \in A$ precisely one $b \in B$. A is called the **domain** of f and B is called the **codomain** of f. When f assigns a to b, we say that $f(a) = b$. We read this 'f of a equals b'.

3.2 Example

Let $f(x) = x + 2 \, \forall \, x \in \mathbb{R}$.

Claim: f is a function from $\mathbb{R}$ to $\mathbb{R}$.

Proof. Let $x \in \mathbb{R}$. $f(x) = x + 2$. $x + 2 \in \mathbb{R}$. And there is only one number in $\mathbb{R}$ that is equal to $x + 2$. So f is a function from $\mathbb{R}$ to $\mathbb{R}$.
$\square$

Showing a function really is a function is often called showing a function is well-defined.

3.3 Notation

Let A and B be sets. If we want to say 'f is a function from A to B', we will often write $f : A \longrightarrow B$. So in 3.2 on the preceding page, $f : \mathbb{R} \longrightarrow \mathbb{R}$.

3.4 Example

A function $f : A \longrightarrow B$ can assign lots of different elements of A to the same element of B. For example, let $f : \mathbb{Z} \longrightarrow \{0, 1, 2, 3\}$, where $f(x) = 1 \; \forall \, x \in \mathbb{Z}$. f really is a function. Every element of $\mathbb{Z}$ gets assigned to exactly one element in $\{0, 1, 2, 3\}$. And lots of integers are being assigned to the same thing. Some functions assign every element of the domain to a different element of the codomain ...

3.5 Definition

Let A and B be sets, and let $f : A \longrightarrow B$. We say f is **1-1** (read "one to one") or **injective** when $f(x) = f(y)$ implies $x = y$. Or, equivalently, when x $\neq$ y implies that $f(x) \neq f(y)$.

3.6 Example

Let $f : \mathbb{Z} \longrightarrow \mathbb{R}, f(x) = 2x$.

Claim: f is 1-1.

Proof. Suppose $f(x) = f(y)$. $f(x) = 2x$ and $f(y) = 2y$. So $2x = 2y$. So $x = y$. So f is 1-1.

$\square$

3.7 Example

Let $X = \{x \in \mathbb{R}; \ x \geq 0\}$. Let $f : \mathbb{R} \longrightarrow X, f(x) = x^2$.

Claim: f is not 1-1.

Proof. $f(1) = 1 = f(-1)$, but $1 \neq -1$. So f is not 1-1. □

3.8 Remark

In our definition of a function we required every element in the domain to be sent to an element of the codomain. However, we didn't require that every element in the codomain get an element of the domain assigned to it. For example, let $f : \{0, 1, 2\} \longrightarrow \{0, 1, 2\}, f(x) = 0 \ \forall \ x \in \{0, 1, 2\}$. Then f is a function, but not every element of the codomain gets an element of the domain assigned to it (for example, there is no $t \in \{0, 1, 2\}$ such that $f(t) = 1$). Some functions assign an element of the domain to every element of the codomain.

3.9 Definition

Let A, B be sets and $f : A \longrightarrow B$. When $\forall b \in B \ \exists \, a \in A$ such that $f(a) = b$, we say that f is **onto**. Equivalently, we may say f is **surjective**.

3.10 Example

Claim: The function f in 3.2 is onto.

Proof. Let $y \in \mathbb{R}$. Then $y - 2 \in \mathbb{R}$.
And $f(y - 2) = (y - 2) + 2 = y$. So f is onto. (Why are we done?) □

3.11 Example

Claim: The function f in 3.4 is not onto.

Proof. $2 \in \{0, 1, 2, 3\}$. And there is no $x \in \mathbb{Z}$ such that $f(x) = 2$. So f is not onto. □

3.12 Remark

Suppose you have a function $f : A \longrightarrow B$.
If f is 1-1, you might think of A as at least as 'big' as B.
If f is onto, you might think of B as at least as 'big' as A.
See 3.27 on page 35 for an example of why care should be taken in this interpretation.

Some people get excited about functions that are both injective and surjective.

3.13 Definition

Let $f : A \longrightarrow B$. Let f be injective and surjective. Then we say that f is **bijective** . We call f a **bijection** from A to B or from A onto B.

3.14 Example

Let $f : \{0, 1\} \longrightarrow \{0, 1\}$ where $f(0) = 1$ and $f(1) = 0$.

Claim: f is bijective.

Proof. f is both 1-1 and onto. □

Part I: Preliminary Material

3.15 Example

Let $f : \mathbb{Q} \longrightarrow \mathbb{R}$, $f(x) = x$.

Claim: f is 1-1, but not bijective.

Proof. f is 1-1:
Suppose $f(x) = f(y)$. $f(x) = x$ and $f(y) = y$. So $x = y$.

f is not onto:
$\sqrt{2} \in \mathbb{R}$. But there is no $x \in \mathbb{Q}$ such that $f(x) = \sqrt{2}$. Why? Suppose that there is such an x. Then $f(x) = x = \sqrt{2}$. But we claim that $\sqrt{2} \notin \mathbb{Q}$.

Lemma: $\sqrt{2} \notin \mathbb{Q}$

Proof of Lemma. Suppose that $\sqrt{2} \in \mathbb{Q}$. Then for some $x, y \in \mathbb{Z}$, $\sqrt{2} = \frac{x}{y}$, where $x \neq 0$ and $\frac{x}{y}$ is written in lowest terms (that is, x and y are relatively prime).
So $\sqrt{2}^2 = \frac{x^2}{y^2}$.
So $2 = \frac{x^2}{y^2}$.
So $2y^2 = x^2$. The square of an odd number is odd, and $2y^2$ is even. So x must not be odd. That is, x is even.
So $x = 2t$ for some $t \in \mathbb{Z}$. So $2y^2 = x^2 = (2t)^2 = 4t^2$.
So we have $2y^2 = 4t^2$.
So $y^2 = 2t^2$.
Since the square of an odd number is odd, and since y^2 is even, y must be not odd. That is, y is even.
So x and y are even. But we assumed that x and y are relatively prime. So this is a contradiction to our assumption that $\sqrt{2} \in \mathbb{Q}$. So $\sqrt{2} \notin \mathbb{Q}$. □

Our assumption that $\exists x \in \mathbb{Q}$ such that $f(x) = \sqrt{2}$ has led to a contradiction. So there is no such x. So f is not onto. □

3.16 Definition

Let A, B, and C be sets and $f : A \longrightarrow B$ and $g : B \longrightarrow C$. Then we can define $h = g \circ f$, $h : A \longrightarrow C$ with $h(x) = g(f(x))\ \forall\ x \in A$. That is, we take $x \in A$, we use f to send it to $f(x)$ which is an element of B. Then we take this element of B, $f(x)$, and we apply g to it getting $g(f(x))$. We say h is g **composed** with f or the **composition** of g with f.

3.17 Example

Let $f : \mathbb{Z} \longrightarrow \mathbb{Q}$, $g : \mathbb{Q} \longrightarrow X$ where $X = \{x \in \mathbb{R}; x \geq 0\}$.
Let $f(x) = x\ \forall\ x \in \mathbb{Z}$. And let $g(x) = \sqrt{x}\ \forall\ x \in \mathbb{Q}$.
Let $h = g \circ f$.
Then $h(2) = g(f(2)) = g(2) = \sqrt{2}$.
$h(16) = g(f(16)) = g(16) = \sqrt{16} = 4$.

3.18 Definition

Let $f : A \longrightarrow A$. When $f(x) = x\ \forall\ x \in A$, we say that f is the **identity** on A. We will sometimes write $f = id_A$ or $f = id : A \longrightarrow A$.

3.19 Definition

Let $f : X \longrightarrow Y$ and $g : X \longrightarrow Y$. We say that $f = g$ when $f(x) = g(x)\ \forall\ x \in X$.

3.20 Definition

Let $f: A \longrightarrow B$, $g: B \longrightarrow A$. Let $h = g \circ f$, and $j = f \circ g$. Note that $h: A \longrightarrow A$ and $j: B \longrightarrow B$. When $h = id_A$ and $j = id_B$, we say that f is the **inverse** of g, and g is the inverse of f. That is, $f = g^{-1}$. The use of the phrase 'the inverse of' is ok, because the inverse of a function is unique.

Note that if $f = g^{-1}$, then we automatically have $g = f^{-1}$.

We say that f and g are inverses of each other.

3.21 Result

Let $id: X \longrightarrow X$.

Claim: $id^{-1} = id$

Proof. $id(id(x)) = id(x) = x \; \forall \; x \in X$. So id is its own inverse. $\square$

3.22 Example

Let $A = \{x \in \mathbb{R}; \; x \geq 0\}$
Let $f: A \longrightarrow A, f(x) = \sqrt{x} \; \forall \; x \in \mathbb{R}$.
Let $g: A \longrightarrow A, g(x) = x^2 \; \forall \; x \in \mathbb{R}$.

Claim: $f = g^{-1}$

Proof. $f(g(x)) = f(x^2) = \sqrt{(x^2)} = x$
$g(f(x)) = g(\sqrt{x}) = (\sqrt{x})^2 = x$ $\square$

3.23 Result

Let $f : X \longrightarrow Y$.

Claim: f^{-1} exists if and only if f is bijective.

Proof. $\Longrightarrow$ Suppose $\exists\, f^{-1} : Y \longrightarrow X$.
f is onto:
Let $y \in Y$. Then $f^{-1}(y) = x$ for some $x \in X$, since f^{-1} is well-defined. And $f(x) = f(f^{-1}(y)) = (f \circ f^{-1})(y) = id_Y(y) = y$.

f is 1-1:
Let $f(x_1) = f(x_2)$
Then $f^{-1}(f(x_1)) = f^{-1}(f((x_2))$
$(f^{-1} \circ f)(x_1) = (f^{-1} \circ f)(x_2)$
But $f^{-1} \circ f = id_X$.
So we have $id(x_1) = id(x_2)$.
So $x_1 = x_2$.
So we now have f is bijective.

$\Longleftarrow$
Let $f : X \longrightarrow Y$ be a bijection.
We want to construct f^{-1}. Let $y \in Y$. Then $\exists\, x \in X$ such that $f(x) = y$, since f is onto. Since f is 1-1, we know this x is unique. So for each y there is exactly one x such that $f(x) = y$. And that's good.

For each $y \in Y$, choose the one $x \in X$ such that $f(x) = y$, and let $f^{-1}(y) = x$. We have defined the function f^{-1}. So we are done. □

Part I: Preliminary Material

3.24 Problem

Let X, Y, and Z be sets.
Let $f : X \longrightarrow Y$ and $g : Y \longrightarrow Z$ be bijective and define $h = g \circ f$. Show h^{-1} exists and $h^{-1} = f^{-1} \circ g^{-1}$.

3.25 Problem

Let $f : A \longrightarrow B$, $g : B \longrightarrow C$, $h = g \circ f$.
Decide whether each of the following four statements is true or false and show why.
1) If h is surjective, then f is surjective.
2) If h is surjective, then g is surjective.
3) If h is injective, then f is injective.
4) If h is injective, then g is injective.

3.26 Problem

Let $f : A \longrightarrow B$, $g : B \longrightarrow C$, $h = g \circ f$.
Show the following:
1) If f and g are injective, then h is injective.
2) If f and g are surjective, then h is surjective.
3) If f and g are bijective, then h is bijective.

3.27 Problem

Construct a bijection from $[0, 1]$ to $(0, 1)$.
Hint: Let $A = \{\frac{1}{2}, \frac{1}{3}, \frac{1}{4}, \frac{1}{5}, \ldots\}$ and let $B = \{x; x \in (0, 1) - A\}$. Construct a function f where $f(b) = b \ \forall \ b \in B$, and $f(\frac{1}{n}) = \frac{1}{n+2} \ \forall \ n \in \{2, 3, 4, 5, \ldots\}$. You should then define $f(0)$ and $f(1)$ in a useful way, and show that f is indeed a

bijection.

This gives a reason why we should be careful using the word "big" for infinite sets in the manner it is used in 3.12 on page 30. Why?

3.28 Problem

Let $f : A \longrightarrow A$, $g = f \circ f$, and g injective. Show f is injective.

3.29 Problem

Let $id : X \longrightarrow X$. Show id is bijective.

3.30 Definition

Let $A \subset X, f : X \longrightarrow Y$. Define $f(A) = \{y \in Y;\ y = f(x)$ for some $x \in A\}$. We sometimes call $f(A)$ the **image** of A under f.

Let $B \subset Y$. Define $f^{-1}(B) = \{x \in X;\ f(x) \in B\}$. We sometimes call $f^{-1}(B)$ the **inverse image** of B under f.

Note that $f^{-1}(B)$ exists even when no function f^{-1} exists.

3.31 Example

Define $f : \mathbb{R} \longrightarrow \{-1, 1\}$, $f(x) = \begin{cases} -1 & \text{for } x \leq 0 \\ 1 & \text{for } x > 0 \end{cases}$

Note this is read as $f(x)$ is -1 for x less than or equal to 0, and $f(x)$ is 1 for x greater than 0.

$f([-2, 1]) = \{-1\}$

$f([-1, 0.1)) = \{-1, 1\}$

$f^{-1}(\{-1\}) = (-\infty, 0]$

$f^{-1}(\{1\}) = (0, \infty)$

Chapter 4

Exam

It is strongly recommended that you do not continue to the next part until the following exam is easy.

1)
Define $x \in \mathbb{Z}$ to be odd if $x = 2t + 1$ for some $t \in \mathbb{Z}$.
Define $y \in \mathbb{Z}$ to be even if $x = 2t$ for some $t \in \mathbb{Z}$.
Show all of the following:
There is no integer that is both even and odd.
The sum of two odd integers is even.
The sum of two even integers is even.
The sum of an odd integer and an even integer is odd.
The square of an odd integer is odd.
The square of an even integer is even.

2) Let $A \subset B$. Show that $A \bigcap B = A$ and $A \bigcup B = B$.

3) Let X be a set. Show that $\phi \bigcap X = \phi$ and $\phi \bigcup X = X$.

4) State the negations of the following two statements:
a) There are no x's in Y such that X is blarg but not bloog.
b) All x's are bloog.

5) Define $X = \{0, \{0\}, 1\}$. Which of these are elements of X
a) $\{1\}$
b) ϕ
c) X

6) Let f be defined as in 3.15 on page 31. We showed f is not bijective. Which of the following exist
a) $f^{-1}(\mathbb{Q})$
b) $f^{-1}(\mathbb{R} - \mathbb{Q})$
c) $f^{-1}(\{2\})$
d) $f^{-1}(\{\sqrt{2}\}$

Part II

Definition of Topology

Chapter 1

Topology

1.1 Definition

Let X be a set. A **topology** on X is a set T whose elements are subsets of X having the following three properties:

1) ϕ and X are open
2) Given any collection of open sets, their union is open
3) Given any two open sets, their intersection is open

Each element of T called an **open** set.

When we have a set X and a topology T on X, we call (X, T) a **topological space** or simply a **space**. We say that X is equipped with the topology T. When it should cause no confusion, we will sometimes just say X is a space (omitting naming the topology).

The definition of a topology may seem a bit awkward. We will do lots of examples to try to make dealing with it easy.

1.2 Notation

Let (X, T) be a space.

The following all mean the same thing:
1) U is an open set in the topological space (X, T)
2) $U \in T$

And, if the topology T is understood:
3) U is open in X
4) $U^{open} \subset X$

And, if both the set X and the topology T on it are understood:
5) U is open

1.3 Example

Let X be a set. Let $T = \{\phi, X\}$

Claim: (X, T) is a topological space.

Proof. 1) $\phi, X \in T$.

2) A collection of open sets either includes X or does not. If it includes X, then the union of this collection is X, which is open. If it does not include X, then every element in the collection is ϕ. And the the union is ϕ, which is open. So for any collection of open sets, we have shown that their union is open.

3) Let $U_1, U_2 \in T$.

Part II: Definition of Topology   45

Case 1: U_1 or $U_2 = \phi$
Then $U_1 \cap U_2 = \phi$, which is open.

Case 2: U_1 and $U_2 \neq \phi$.
Then $U_1 = U_2 = X$. And $U_1 \cap U_2 = X$, which is open.

So (X, T) is a topological space. □

1.4 Definition

When a set X is equipped with a topology T and the elements of T are precisely ϕ and X (as in 1.3), T is said to be the **indiscrete topology**.

Note that we have just found a way to turn any set into a topological space. The set X might have numbers, or cars, or matrices as elements. It doesn't matter. We now have a way to turn the set into a space.

1.5 Example

Let X be a set, and let $T = \{U; U \subset X\}$.
In other words, let every subset of X be open in X.

Claim: T is a topology on X.

Proof. 1) $\phi \subset X$, so ϕ is open.
$X \subset X$, so X is open.

2) Let $\{X_\alpha\}$ be a collection of open sets of X. Each X_α is contained in X. So the union of these sets is contained in X, and thus open in X.

3) Let $U_1, U_2^{open} \subset X$. $U_1 \cap U_2 \subset X$. So $U_1 \cap U_2$ is open in X.

$\square$

1.6 Definition

When a set X is equipped with a topology T and every subset of X is an element of T (as in 1.5), then we say that X has the **discrete topology**.

1.7 Example

Let $X = \{0, 1\}$. Let $T = \{\phi, \{0, 1\}, \{0\}, \{1\}\}$.
T is the discrete topology on X. Or equivalently, (X, T) is discrete.

1.8 Definition

Let (X, T) be a space. When $x \in U^{open} \subset X$ (that is, $x \in U$ and $U^{open} \subset X$), then we say U is a **neighborhood** of x.

1.9 Example

Let $X = \{0, 1, 2\}$. Let $T = \{\phi, \{0, 1, 2\}, \{0, 1\}, \{1, 2\}\}$.

Claim: T is not a topology on X.

Proof. $\{0, 1\}$ and $\{1, 2\}$ are open. $\{0, 1\} \cap \{1, 2\} = \{1\}$ which is not open. So (X, T) is not a topological space. $\square$

Part II: Definition of Topology

1.10 Problem

Let $X = \{0, 1, 2\}$. Let $T = \{\phi, \{0, 1, 2\}, \{0, 1\}, \{1, 2\}, \{1\}\}$. Show (X, T) is a space.

1.11 Definition

Let (X, T) be a space. Saying x is a **point** of X is equivalent to saying $x \in X$.

1.12 Problem

Let $X = \{\text{Jason, Luke}\}$ How many different topologies are there on X? Suppose we take a set Y that has two elements, but the elements are not necessarily Jason and Luke. How many different topologies can you place on that set?

1.13 Problem

Let $X = \{\text{the greek letter } \pi, \text{ the number represented by the greek letter } \pi, 3.1\}$. How many different topologies can be placed on X?

1.14 Problem

In $\mathbb{R}$, for $a < b$, an **open interval** (a, b) is the set of real numbers greater than a and less than b. In other words, $(a, b) = \{x;\ a < x < b\}$. We require $a < b$. If $a \geq b$, then (a, b) is not defined.

Suppose you try to define a topological space $(\mathbb{R}, T)$ where the open sets are ϕ, $\mathbb{R}$ and every open interval (a, b) where

$a, b \in \mathbb{R}$? Is $(\mathbb{R}, T)$ is a topological space? (Hint: the answer is no).

Chapter 2

Standard Topology on $\mathbb{R}$

2.1 Definition

Recall that in 1.14 on the facing page we defined an open interval in $\mathbb{R}$.

Let $T = \{A;\ A = \phi$ or A is a union of open intervals in $\mathbb{R}\}$ Then we call T the **standard topology on** $\mathbb{R}$. We call it a topology because it is one. But we need to show that it is, which we do now ...

2.2 Result

Let T be the standard topology on $\mathbb{R}$.

Claim: $(\mathbb{R}, T)$ is a topological space.

Proof. 1) Unions of open sets are unions of unions of open intervals which are unions of open intervals, thus open.

2) ϕ is open. And for each $n \in \mathbb{N}$, $(-n, n)$ is open. So $\bigcup_{n=1}^{\infty}(-n, n)$ is open (by 1) above). It would be nice if $\bigcup_{n=1}^{\infty}(-n, n) = \mathbb{R}$. Then we will have shown that $\mathbb{R}$ is open.

Lemma: $\bigcup_{n=1}^{\infty}(-n,n) = \mathbb{R}$

Proof of Lemma. LHS $\subset$ RHS:
Let $x \in \bigcup_{n=1}^{\infty}(-n,n)$. Then $x \in (-n,n)$ for some $n \in \mathbb{N}$. So $x \in \mathbb{R}$.
So $x \in$ RHS.

RHS $\subset$ LHS:
Let $x \in$ RHS. Then $x = 0, x > 0$, or $x < 0$.

Case 1: If $x = 0$, then $x \in (-1,1)$ and so $x \in$ LHS.

Case 2: If $x > 0$, then $\exists\, t \in \mathbb{N}$ such that $t > x$. And $x \in (-t,t)$. So $x \in$ LHS.

Case 3: If $x < 0$, then $\exists\, t \in \mathbb{N}$ such that $-t < x$. And $x \in (-t,t)$. So $x \in$ LHS.

So we have shown that RHS $\subset$ LHS.
So we have shown that LHS = RHS.
So we are done with both the lemma and 2). □

3) Let U_1, U_2 be open. Each is a union of open intervals or empty. $U_1 \cap U_2$ is a union of open intervals or empty (Why?). So $U_1 \cap U_2$ is open.

So $(\mathbb{R}, T)$ is a topological space. □

2.3 Example

Let $\mathbb{R}$ have the standard topology. Let $a \in \mathbb{R}$.
Let $(a, \infty) = \{x \in \mathbb{R};\ x > a\}$.

Claim: (a, ∞) is open.

Part II: Definition of Topology

Proof. There are lots of integers bigger than a. Choose one, and call it T. Look at (a, T), $(a, T+1)$, $(a, T+2)$, etc. Let $A_i = (a, T+i) \, \forall \, i \in N \bigcup \{0\}$.
Each A_i is open. So $\bigcup_{i=0}^{\infty} A_i$ is open.
If we can show that $\bigcup_{i=0}^{\infty} A_i = (a, \infty)$, then we are done.

Lemma: $\bigcup_{i=0}^{\infty} A_i = (a, \infty)$

Proof of Lemma. Let $x \in \bigcup_{i=0}^{\infty} A_i$. Then $x \in A_i$, for some $i \in \{0, 1, 2, \ldots\}$. So $x \in (a, y)$ for some $y \in \mathbb{R}$, $y > a$. So $x \in (a, \infty)$. Let $x \in (a, \infty)$. $\exists \, M, T \in \mathbb{N}$ such that $M > T > x$. $x \in (a, M) = A_{M-T}$. (Why?)
So $x \in \bigcup_{i=0}^{\infty} A_i$.

□

□

2.4 Problem

Let $\mathbb{R}$ have the standard topology. Show $(-\infty, a)$ is open in $\mathbb{R}$.

2.5 Remark

Let $a, b \in \mathbb{R}, a > b$.
$[a, b] = \{x \in \mathbb{R}; \, a \leq x \leq b\}$.
$[a, b) = \{x \in \mathbb{R}; \, a \leq x < b\}$.
$(a, b] = \{x \in \mathbb{R}; \, a < x \leq b\}$.

$[a, b], [a, b), (a, b]$ are all not open in $\mathbb{R}$. It is awkward to show this with the tools we have now. We would have to show that we could not write any of them as a union of open intervals. In Chapter 5 we will introduce some tools that make showing these are not open easy.

2.6 Remark

Note that we have some topologies that we can place on any set (discrete, indiscrete), and one topology (the standard topology on $\mathbb{R}$) that we can only place on $\mathbb{R}$. In what sense does it make sense to think of the discrete and indiscrete topologies on a given set as two extremes?

Next we will introduce one more topology we can place on any set. Then we will introduce a little more topological language and apply it a bunch to all of the topologies that we have discussed.

Chapter 3

Cofinite Topology

3.1 Definition

Let X be a set, and let $A \subset X$. When $X - A$ is finite, we say A is **cofinite**.

3.2 Result

Let X be a set and let $T = \{U \subset X;\ U \text{ is cofinite or } U = \phi\}$

Claim: T is a topology on X.

Proof. 1) ϕ is open.
$X - X = \phi$ is finite, so X is open.

2) Take a collection of open sets. We know that each set is empty or has finite complement. Let's take the union of all the open sets. We hope this is open.

Case 1: All the sets are empty.
Then the union of all the sets is empty, which is open.

Case 2: Not all the sets are empty.
We want to show the union is open. We will show the complement of the union is finite. The complement of the union is the intersection of the complements, by 2.30 on page 24. We know at least one of the sets is not empty. So the complement of at least one of the sets is finite. So the intersection of all the complements is finite. So the complement of the union is finite. So the union is open.

3) Let U_1, U_2 be open. $X - (U_1 \cap U_2) = (X - U_1) \cup (X - U_2)$ (by 2.31 on page 24)

Case 1: $X - U_1 = X$ or $X - U_2 = X$.
Then $(X - U_1) \cup (X - U_2) = X$ which is open. So $U_1 \cap U_2$ is open.

Case 2: $X - U_1 \neq X$ and $X - U_2 \neq X$.
Then $X - U_1$ and $X - U_2$ are finite.
The union of two finite sets is finite.
So $X - (U_1 \cap U_2)$ is finite.
So $U_1 \cap U_2$ is open.

So (X, T) is a topological space.

□

3.3 Definition

Let X be any set. When we equip X with the topology $T = \{U \subset X; U \text{ is cofinite or } U = \phi\}$ we say X is equipped with the **cofinite topology** .

Part II: Definition of Topology   55

3.4 Example

Suppose X is finite, and let T be the cofinite topology on X. What are the open sets of the space (X, T)? For every subset of X, the complement of it is also a subset of X. So the complement of it is finite (since every subset of a finite set is finite). So the complement of every subset of X is finite. So every subset of X is open. So, what have we shown? We've shown that if X is finite, the cofinite topology on X is the same as the discrete topology on X.

3.5 Example

Equip $\mathbb{R}$ with the cofinite topology. Let's figure out what some of the open sets look like.

We know ϕ and $\mathbb{R}$ are open, so there's two open sets.
The complement of $(-\infty, 0) \bigcup (0, \infty)$ is $\{0\}$ which is finite. So $(-\infty, 0) \bigcup (0, \infty)$ is open. In fact, for every $a \in \mathbb{R}$, the complement of $(-\infty, a) \bigcup (a, \infty) = \{a\}$ which is finite. So $(-\infty, a) \bigcup (a, \infty)$ is open for every $a \in \mathbb{R}$. So we've now found infinitely many open sets in $(\mathbb{R}, T)$. Let's find more.

Can you find an open set that we haven't mentioned yet? There are infinitely many that we haven't mentioned.

Now let's find some sets that are not open. Let's prove a lemma first ...

Lemma: Let Y be finite, $Y \subset X$, and X infinite. Then $X - Y$ is infinite.

Proof of Lemma. $Y \bigcup (X - Y) = X$. Suppose $X - Y$ were finite. Then $Y \bigcup (X - Y)$ would be finite (since Y is also finite). But $Y \cup (X - Y) = X$ and X is not finite. So $X - Y$ cannot be finite. So $X - Y$ is infinite.

$\square$

The complement of any non-empty finite set in $\mathbb{R}$ is infinite. (by Lemma)

So every non-empty finite set in $\mathbb{R}$ is not open in the cofinite topology. Why is the word 'non-empty' necessary in the previous sentence?

So we now have infinitely many sets that are not open. For example, $\{1\}, \{1, 2\}, \{1, 2, 3\}$, etc ... are not open.

3.6 Problem

Let $\mathbb{R}$ have the cofinite topology. Show that there are infinitely many infinite sets that are not open in $\mathbb{R}$.

3.7 Problem

Let $X = \{0, 1, 2, \ldots, 100019871298719865652^{2183217681586512}\}$. Equip X with the finite complement topology. What are the open sets of X? How does the answer change if we make 2183217681586512 a bigger positive integer?

If $Y = \{a, b\}$ and we equip Y with the cofinite topology, then what are the open sets in Y? List them explicitly.

3.8 Result

Claim: On $\mathbb{R}$ the cofinite topology is not the same as the discrete topology.

Proof. In the cofinite topology, [0, 1] is not open. In the discrete topology, [0, 1] is open. □

3.9 Result

Claim: On $\mathbb{R}$ the cofinite topology is not the same as the indiscrete topology.

Proof. In the cofinite topology, $(-\infty, 0) \bigcup (0, \infty)$ is open. In the indiscrete topology, it is not. □

3.10 Result

Claim: On $\mathbb{R}$ the cofinite topology is not the same as the standard topology.

Proof. In the cofinite topology $(0, 1)$ is not open. In the standard topology it is. □

Chapter 4

Closed Sets

4.1 Definition

Let $A \subset X$. A is said to be **closed** in X when $X - A$ is open in X. We may sometimes write $A^{closed} \subset X$. We will sometimes simply say A is closed when its being closed in X should be clear.

4.2 Result

Let (X, T) be a space.

Claim: X is both open and closed in X.

Proof. X must of course be open in X since X is a space. To show X is closed, we need to show that $X - X$ is open. But $X - X = \phi$ which is open. So X is closed in X. □

4.3 Result

Let (X, T) be a space.

Claim: ϕ is both open and closed in X.

Proof. ϕ is open in X, since X is a space. To show ϕ is closed in X, we need to show that $X - \phi$ is open. But $X - \phi = X$ which is open. So ϕ is closed in X. □

4.4 Result

Let X have the cofinite topology.

Claim: $A^{closed} \subset X \iff A$ is finite or $A = X$.

Proof. $\implies$ Let $A^{closed} \subset X$. Then $X - A$ is open in X. So A is finite or $A = X$.

$\impliedby$ Suppose A is finite or $A = X$.
Case 1: A is finite. Then $X - A$ is open. So A is closed.
Case 2: $A = X$. Then A is closed by 4.2. □

4.5 Example

Equip $\mathbb{R}$ with the standard topology.

Claim: $[0, 1]$ is closed in $\mathbb{R}$.

Proof. $\mathbb{R} - [0, 1] = (-\infty, 0) \bigcup (1, \infty)$. This is a union of open sets (by 2.3 on page 50 and 2.4 on page 51), thus open. So $[0, 1]$ is closed. □

4.6 Example

Equip $\mathbb{R}$ with the standard topology.

Part II: Definition of Topology   61

Claim: $\{a\}$ is closed for any $a \in \mathbb{R}$.

Proof. $\mathbb{R} - \{a\} = (-\infty, a) \bigcup (a, \infty)$. This is the union of open sets (by 2.3 on page 50 and 2.4 on page 51) thus open. So $\{a\}$ is closed. □

4.7 Example

Equip $\mathbb{R}$ with the standard topology.

Claim: Every finite set in $\mathbb{R}$ is closed.

Proof. Let A be a finite set in $\mathbb{R}$. Let the finite number of (distinct) elements of A be $x_1, x_2, \ldots, x_n$. Order these real numbers from smallest to greatest, $a_1 < a_2 < \ldots < a_n$, where a_1 is the smallest of the $x_{i's}$, a_2 is the second smallest of the $x_{i's}$, ..., and a_n is the largest of the $x_{i's}$.

Let $B = \{a_1, a_2, \ldots, a_n\}$. Note that $B = A$ (we have just reordered the elements of A in a convenient way).

$\mathbb{R} - B = (-\infty, a_1) \bigcup (a_1, a_2) \bigcup \ldots \bigcup (a_{n-1}, a_n) \bigcup (a_n, \infty)$. This is a union of open sets in $\mathbb{R}$, thus open. So B is closed. And A, which is equal to B, is also closed. □

4.8 Example

Equip $\mathbb{R}$ with the standard topology. Let $a, b \in \mathbb{R}, a < b$.

Claim: $[a, b]$ is closed in $\mathbb{R}$.

Proof. $\mathbb{R} - [a, b] = (-\infty, a) \bigcup (b, \infty)$. This is the union of open sets in $\mathbb{R}$, thus open. So $[a, b]$ is closed in $\mathbb{R} \forall a, b \in \mathbb{R}$. □

4.9 Problem

Let X be a space. Let $A^{open} \subset X$, $B^{closed} \subset X$, $B \subset A$. Show $A - B$ is open. (Hint: Show $A - B = A \cap (X - B)$)

4.10 Result

Let X be any set and equip it with the discrete topology.

Claim: Every subset of X is closed.

Proof. Let $U \subset X$. $X - U \subset X$, so $X - U$ is open. So U is closed. $\square$

4.11 Result

Let X be any set. Let X have the indiscrete topology.

Claim: The only closed sets in X are ϕ and X.

Proof. Suppose $U^{closed} \subset X$. Then $X - U$ is open. The only open sets in X are ϕ and X. So $X - U = \phi$ or $X - U = X$. If $X - U = \phi$, then $U = X$. If $X - U = X$, then $U = \phi$. $\square$

4.12 Result

Let $X = (0,1) \cup [2,3]$. Let $\mathbb{R}$ have the standard topology. Let $T = \{V; \ V = U \cap X \text{ for some } U^{open} \subset \mathbb{R}\}$ (in 1.1 on page 113 we will show that (X, T) is indeed a space, which we will call the 'subspace topology').

Claim: $[2,3]$ is both open and closed in (X, T).

Part II: Definition of Topology

Proof. $[2,3] = (1.5, 3.5) \cap X$. And $(1.5, 2.5)$ is open in the standard topology on $\mathbb{R}$. So $[2,3]^{open} \subset X$.

To show $[2,3]$ is closed in X, we need to show that $X - [2,3]$ is open in X. $X - [2,3] = (0, 1)$. We need to show that $(0, 1)$ is open in X.

$(0, 1) = (0, 1) \cap X$. And since $(0, 1)$ is open in the standard topology on $\mathbb{R}$, we have shown that $(0, 1)$ is open in X. So $[2, 3]$ is closed in X. $\square$

4.13 Result

Let $\mathbb{R}$ have the cofinite topology and define
$A = (-\infty, 0) \cup (0.000001, \infty)$

Claim: A is not open in $\mathbb{R}$ and A is not closed in $\mathbb{R}$.

Proof. $R - A = [0, 0.000001]$ which is not finite and not empty. So A is not open in $\mathbb{R}$. A is not finite, and A does not equal $\mathbb{R}$. So, by 4.4, A is not closed. $\square$

Chapter 5

Some Useful Tools

5.1 Definition

Let (X, T) be a space. Let $x \in A \subset X$. x is said to be an **interior point** of A if $\exists$ a neighborhood U of x such that $U \subset A$.

5.2 Example

Look at $\mathbb{R}$ with the standard topology. Let $A = (2.3, 8) \subset \mathbb{R}$.

Claim: 5 is an interior point of A and 2.3 is not an interior point of A.

Proof. $5 \in (4, 6)$ which is open in $\mathbb{R}$ and contained in A. So 5 is an interior point of A.

Suppose U is any neighborhood of 2.3. Then U is a union of open intervals in $\mathbb{R}$ (since U is open and these are the only non-empty open sets). 2.3 is an element of at least one of these open intervals. Choose one of them and call it (a, b). Since $2.3 \in (a, b)$, $2.3 \neq a$ and $2.3 \neq b$. So $a < 2.3$. So (a, b)

is not contained in A. (Why? ... (a,b) has lots of points in it that are not in A. For example, the number $\frac{(2.3+a)}{2}$ is one of them.)

And since $(a,b) \subset U$, U is not contained in A. So what have we done? Given any neighborhood of 2.3, we have shown that this neighborhood is not contained in A. So we have shown that 2.3 is not an interior point of A. □

5.3 Definition

Let $x \in A \subset X$. x is said to be a **limit point** of A if every neighborhood of x intersects $A - \{x\}$. Equivalently, x is a limit point of A if the following is true: $x \in U^{open} \subset X \implies U \cap (A - \{x\}) \neq \phi$.

5.4 Problem

Look at $\mathbb{R}$ with the standard topology. Let $A = (2.3, 8) \subset \mathbb{R}$. Show 5 and 2.3 are both limit points of A.

5.5 Result

Let X be a space, $A \subset X$.

Claim: $A^{open} \subset X \iff$ every element of A is an interior point of A.

Proof. $\implies$ Let A be open. Let $x \in A$. Then since $A \subset A$, A is a neighborhood of x in A. So x is an interior point.

$\impliedby$ Suppose every element of A is an interior point of A. Then for each $x \in A$ we have a neighborhood of x, $U_x \subset A$.

Part II: Definition of Topology   67

Take the union of all these neighborhoods, $\bigcup_{x \in X} U_x$.

Lemma: $\bigcup_{x \in X} U_x = A$

Proof of Lemma. Let $x \in \bigcup_{x \in X} U_x$. Then $x \in U$ for some U that is one of the sets in the union. $U \subset A$. So $x \in A$. Let $x \in A$. Then $x \in U_x$ and $U_x \subset \bigcup_{x \in X} U_x$. So $x \in \bigcup_{x \in X} U_x$. □

$\bigcup_{x \in X} U_x$ is open, since it is a union of open sets. So, by the Lemma, A is open. □

5.6 Result

Claim: $A^{closed} \subset X \iff A$ contains all its limit points.

Proof. $\implies$ Let A be closed. Let $x \in X - A$. We want to show that x is not a limit point of A. Since A is closed, $X - A$ is open. Let $U = X - A$. $x \in U$. And $U \cap A = \phi$. So we have shown that there is a neighborhood of x, U, that does not intersect A. So x is not a limit point of A. So we have shown that if A is closed, then no point outside of A can be a limit point of A. In other words, A contains all its limit points.

$\impliedby$ Let A contain all its limit points. Let $x \in X - A$. x is not a limit point of A. So $\exists$ a neighborhood of x, U, such that U does not intersect A. So x is an interior point of $X - A$. So, by 5.5, $X - A$ is open. So A is closed. □

5.7 Result

Let $\mathbb{R}$ have the standard topology, let $a, b \in \mathbb{R}$, $a < b$.

Claim: $[a, b]$ is not open in $\mathbb{R}$.

Proof. a is not an interior point of $[a, b]$, because every neighborhood of a slips outside of $[a, b]$ (Why?). That is, for any neighborhood U of a, U is not contained in $[a, b]$. So by 5.5, $[a, b]$ is not open in $\mathbb{R}$. □

5.8 Problem

Let $\mathbb{R}$ have the standard topology, let $a, b \in \mathbb{R}$, $a < b$.
Show $(a, b]$ is not open in $\mathbb{R}$.

5.9 Problem

Let $\mathbb{R}$ have the standard topology, let $a, b \in \mathbb{R}$, $a < b$.
Show $[a, b)$ is not open in $\mathbb{R}$.

5.10 Problem

Let $\mathbb{R}$ have the standard topology.
Find the following:
1) A set that is open and not closed
2) A set that is closed and not open
3) A set that is open and closed
4) A set that is not open and not closed

Chapter 6

Some Notes on the Definition of a Topology

6.1 Result

Let (X, T) be a topological space.

Claim:
1) ϕ and X are closed.

2) The union of any two closed sets is closed.

3) The intersection of any collection of closed sets is closed.

Proof. 1) 4.2 on page 59, 4.3 on page 59

2) Let U_1 and U_2 be closed in X. Then $X - U_1$ is open and $X - U_2$ is open.
So $(X - U_1) \bigcap (X - U_2)$ is open.
But $(X - U_1) \bigcap (X - U_2) = X - (U_1 \bigcup U_2)$.
Since this is open, $U_1 \bigcup U_2$ is closed.

3) Suppose we have a bunch of closed sets. We want to show that their intersection is closed. So we want to show that the complement of their intersection is open. But the complement of their intersection is the union of the complements (by DeMorgan). Since each set is closed, each of the complements is open. And the union of these open sets is open. So we have shown that the complement of their intersection is open. So we have shown that their intersection is closed. So we are done. □

6.2 Result

Claim: The intersection condition in the definition of a topology (in 1.1 on page 43) is equivalent to requiring the following: The intersection of any finite collection of open sets is open.

Proof. $\Longleftarrow$ Suppose we know the intersection of any finite collection of open sets is open. Then we know in particular that the intersection of any two open sets is open.

$\Longrightarrow$ Suppose that the intersection of any two open sets is open. Let $U_1, U_2, \ldots, U_n$ be open. Then $U_1 \cap U_2$ is open. So $(U_1 \cap U_2) \cap U_3$ is open, since it is the intersection of two open sets. Continuing in this way, we get $(U_1 \cap U_2 \cap \ldots \cap U_{n-1}) \cap U_n$ is open.
But this is precisely $U_1 \cap U_2 \cap \ldots \cap U_n$. □

6.3 Problem

Let X be a space. Show the union of any finite collection of closed sets is closed.

Part III

Homeomorphisms

Chapter 1

Basis for a Topology

1.1 Definition

Let X be a set, and B be a collection of subsets of X. If the elements of B satisfy the following two properties, then we call B a **basis** for a topology on X.

1) $\forall x \in X \; \exists B_x \in B$ such that $x \in B_x$.
2) Whenever $x \in B_1 \cap B_2$ for some $B_1, B_2 \in B$, then $\exists B_3 \in B$ such that $x \in B_3 \subset B_1 \cap B_2$.

Given a set X and a basis B we define the topology T generated by B as follows: $U \subset X$ is open when $\forall x \in U \; \exists B_x \in B$ such that $x \in B_x \subset U$.

1.2 Result

Let X be a set, and let B be a basis.

Claim: The 'topology' we defined in 1.1, with basis B, really is a topology.

Proof. 1) ϕ is open vacuously (if it were not open, then there would exist $x \in \phi$ such that there does not exist a $B_x \in B$ such that $x \in B_x \subset U$. But there does not exist any $x \in \phi$ at all).

X is open: Let $x \in X$. We need to show that $\exists B_x \in B$ such that $x \in B_x \subset X$. The definition of a basis gives us a $B_x \in B$ such that $x \in B_x$. And $B_x \subset X$. So X is open.

2) Suppose you have a collection of open sets. We want to show that their union is open. So we want to show that for every x in their union there is a basis element containing x that is a subset of the union. But this is trivial... Since each of the sets in the union is open, for every x in every one of the open sets, we have a basis element containing x that is contained in the open set ... which is contained in the union of the open sets. So we have nothing else to show.

3) Let U_1, U_2 be open.
We want to show that $U_1 \bigcap U_2$ is open. So, we want to show that for every $x \in U_1 \bigcap U_2$ $\exists$ a basis element B_x such that $x \in B_x \subset U_1 \bigcap U_2$.
So let $x \in U_1 \bigcap U_2$. Since $x \in U_1$ (which is open) $\exists B_1 \in B$ such that $x \in B_1 \subset U_1$. Since $x \in U_2$ which is open, $\exists B_2 \in B$ such that $x \in B_2 \subset U_2$.
So $x \in B_1 \bigcap B_2$, the intersection of two basis elements. Since B is a basis , $\exists B_3 \in B$ such that $x \in B_3 \subset B_1 \bigcap B_2$. And $B_1 \bigcap B_2 \subset U_1 \bigcap U_2$. So $x \in B_3 \subset U_1 \bigcap U_2$. So $U_1 \bigcap U_2$ is open.

So a basis for a topology really does define a topology, in the precise way we have mentioned. $\square$

Part III: Homeomorphisms

1.3 Result

Let X be a set, and B be a basis on X. Let T be the topology generated by B. Let $T' = \{U; U \text{ is the union of elements of } B\}$.

Claim: $T = T'$.

Proof. $T \subset T'$:
Let $U \in T$. Since U is open in the topology generated by the basis B, we know that for every $x \in X$ ∃ a basis element B_x such that $x \in B_x \subset U$.

For each $x \in X$, choose such a basis element B_x.
(Note: For each x there may be lots of choices ... pick any one to be B_x.)

Lemma: $U = \bigcup_{x \in X} B_x$

Proof of Lemma. LHS $\subset$ RHS:
Let $x \in U$. Then $x \in B_x \subset U$. So $x \in \bigcup_{x \in X} B_x$.

RHS $\subset$ LHS: Let $x \in \bigcup_{x \in X} B_x$. Then $x \in B_y$ for some B_y in the union. And $B_y \subset U$. So $x \in U$. □

So, by the Lemma, $U \in T'$. And so $T \subset T'$.

$T' \subset T$:
Let $V \in T'$. V is a union of basis elements. We want to show $V \in T$. So, we want to show $\forall\, x \in V\ \exists\, B_x \in B$ such that $x \in B_x \subset V$. So let $x \in V$. Since V is a union of basis elements, we know that x is in at least one of the basis elements of which V is a union. Choose one of them and call it B_x. Since V equals the union of the basis elements, the union of all those basis elements is contained in V. B_x is

one of the sets in this union, so B_x is contained in the union which is contained in V. So we have found a B_x such that $x \in B_x \subset V$. Since our x was arbitrary, we have shown that $V \in T$.

So $T = T'$. $\square$

1.4 Example

Let B be the set of all open intervals in $\mathbb{R}$.

Claim: B is a basis for a topology on $\mathbb{R}$ that is equivalent to the standard topology on $\mathbb{R}$.

Proof: 1.3 and 2.1 on page 49.

1.5 Result

Let (X, T') be a space and let B' be a basis for T'.

Claim: $B \in B' \implies B \in T'$.

Proof. Let $B \in B'$. We want to show that B is open in X. Let $x \in B$. B is a basis element. $x \in B$. And $B \subset B$. So we are done. B is open.

$\square$

1.6 Result

Let (X, T) be a space. Let D be a collection of open sets in (X, T) such that for any $x \in X$ and any neighborhood U of x, there exists $C \in D$ such that $x \in C \subset U$.

Part III: Homeomorphisms

Claim: D is a basis for T.

Proof. 1) Let $x \in X$. We need to find an element of D that contains x. $X^{open} \subset X$. So $\exists\, C \in D$ such that $x \in C \subset X$.

2) Let $x \in C_1 \cap C_2$, with $C_1, C_2 \in D$. C_1 and C_2 are open (since D is a collection of open sets), so $C_1 \cap C_2$ is open. So $\exists\, C_3 \in D$ such that $x \in C_3 \subset C_1 \cap C_2$.

So we have shown that D is a basis for a topology T' on X. Now we need to show that $T' = T$. That is, the topology generated by D is actually the same as T.

$T \subset T'$:
Let $U \in T$. For each $x \in U$ there exists a basis element $C \in D$ such that $x \in C \subset U$. So $U \in T'$ (the topology generated by D).

$T' \subset T$:
Let $U \in T'$. By 1.3, U is the union of elements of D. But each element of D is open in (X, T), by definition. So the union of these elements is also open in (X, T). So U, which is the union of these elements, is also open in (X, T). □

1.7 Remark

Why do people get excited about a basis? 2.4 on page 80 and 2.5 on page 81 explain one reason ...

Chapter 2

Continuous Functions

2.1 Definition

Let $f : X \longrightarrow Y$. Let (X,T), (Y,T') be spaces. f is said to be **continuous** when the inverse image of every open set in Y is open in X. In other words, when $V^{open} \subset Y \implies f^{-1}(V)^{open} \subset X$.

2.2 Result

Let X be a space.

Claim: $id : X \longrightarrow X$ is continuous.

Proof. Let $U^{open} \subset X$. $id^{-1} = id$. So $id^{-1}(U) = id(U) = U$ which is open in X. So id is continuous. $\square$

2.3 Example

Equip $\mathbb{R}$ with its standard topology.
Define $f : \mathbb{R} \longrightarrow \mathbb{R}$, $f(x) = \begin{cases} 0 & \text{for } x \leq 0 \\ 1 & \text{for } x > 0 \end{cases}$
Claim: f is not continuous.

Proof. $(-1, 1)^{open} \subset \mathbb{R}$. $f^{-1}((-1, 1)) = (-\infty, 0]$. This is not open in $\mathbb{R}$ (since 0 is not an interior point of $(-\infty, 0]$). So f is not continuous. $\square$

2.4 Result

Let $(X, T), (Y, T')$ be spaces. Let B' be a basis for T'. Let $f : X \longrightarrow Y$.

Claim: f is continuous $\iff \forall B \in B'$, $f^{-1}(B)^{open} \subset X$.

Proof. $\Longrightarrow$ Suppose that f is continuous. Let $B \in B'$. By 1.5 on page 76 B is open. Since f is continuous, $f^{-1}(B)$ is open in X.

$\Longleftarrow$ Suppose that $\forall B \in B'$, $f^{-1}(B)^{open} \subset X$. Let $V^{open} \subset Y$. Then by 1.3 on page 76 V is a union of basis elements. So $f^{-1}(V) = f^{-1}(\bigcup_{\alpha \in J} f^{-1}(B_\alpha))$ is open. And that's good. Because now we have shown that $f^{-1}(V)$ is open and we are done. $\square$

2.5 Remark

People get excited about 2.4 because it makes for less work. If we want to show $f : X \longrightarrow Y$ is continuous, we no longer have to show that the inverse image of every open set in Y is open in X. If we have a basis for the topology on Y, then

Part III: Homeomorphisms

we only need to show that the inverse image of every basis element is open in X. This is sometimes easier.

2.6 Example

Let $f : \mathbb{R} \longrightarrow \mathbb{R}, f(x) = 2x$.

Claim: f is continuous.

Proof. We make use of 2.4. The standard topology on $\mathbb{R}$ has as a basis all the open intervals in $\mathbb{R}$. Let (a,b) be an open interval in $\mathbb{R}$. $f^{-1}((a,b)) = (\frac{a}{2}, \frac{b}{2})$, which is open in $\mathbb{R}$. So for every basis element in $\mathbb{R}$, we have shown that f^{-1} of that basis element is open in $\mathbb{R}$. By 2.4, f is continuous. □

2.7 Result

Let $f : X \longrightarrow Y$.

Claim: f is continuous $\iff [U^{closed} \subset Y \implies f^{-1}(U)^{closed} \subset X.]$

Proof. $\implies$ Suppose that f is continuous. Let $U^{closed} \subset Y$. Then $Y - U$ is open in Y. And $f^{-1}(Y - U)$ is open in X. We need a lemma ...

Lemma: $f^{-1}(Y - U) = f^{-1}(Y) - f^{-1}(U)$.

Proof of Lemma. RHS $\subset$ LHS:
Let $x \in f^{-1}(Y) - f^{-1}(U)$. Then $f(x) \in Y - U$. So $f(x) \in Y$ and $f(x) \notin U$. So $x \in f^{-1}(Y)$ and $x \notin f^{-1}(U)$. So $x \in f^{-1}(Y) - f^{-1}(U)$.

LHS $\subset$ RHS:
Let $x \in f^{-1}(Y) - f^{-1}(U)$. Then $x \in f^{-1}(Y)$ and $x \notin f^{-1}(U)$. So $f(x) \in Y$ and $f(x) \notin U$. So $f(x) \in Y - U$. So $x \in f^{-1}(Y - U)$.

□

$f^{-1}(Y - U) = f^{-1}(Y) - f^{-1}(U)$, by the Lemma.
And $f^{-1}(Y) = X$.
So $f^{-1}(Y - U) = X - f^{-1}(U)$.
We aleady showed that $f^{-1}(Y - U)$ is open in X. So $f^{-1}(U)$ is closed in X.

$\Longleftarrow$ Suppose $U^{closed} \subset Y \Longrightarrow f^{-1}(U)^{closed} \subset X$. Let $V^{open} \subset Y$. Then $Y - V^{closed} \subset Y$. $f^{-1}(Y - V)^{closed} \subset X$ and $f^{-1}(Y - V) = f^{-1}(Y) - f^{-1}(V) = X - f^{-1}(V)$, applying the above Lemma again. So $f^{-1}(V)$ is open in X. So f is continuous.

□

2.8 Result

Let $f : X \longrightarrow Y$ be continuous and $g : Y \longrightarrow Z$ be continuous. Define $h = g \circ f$.

Claim: $h : X \longrightarrow Z$ is continuous.

Proof. Let $U^{open} \subset Z$. Then $g^{-1}(U)$ is open in Y. For convenience of notation, let $V = g^{-1}(U)$. Since f is continuous, $f^{-1}(V)$ is open in X. $h^{-1} = f^{-1} \circ g^{-1}$. So $h^{-1}(U) = f^{-1}(g^{-1}(U)) = f^{-1}(V)^{open} \subset X$. So h is continuous. □

2.9 Result

Let X be a space.

Claim: X has the discrete topology $\iff \forall$ spaces Y every function $f : X \longrightarrow Y$ is continuous.

Proof. $\Longrightarrow$ Suppose that X has the discrete topology. Let Y be any space and let $f : X \longrightarrow Y$. We want to show that f is continuous. So, choose $V^{open} \subset Y$. $f^{-1}(V) \subset X$. Since X has the discrete topology, $f^{-1}(V)$ is open. So f is continuous.

$\Longleftarrow$ Suppose that $\forall$ spaces Y every function $f : X \longrightarrow Y$ is continuous. We want to show that X has the discrete topology. So we want to show that every subset of X is open. Suppose $\exists U \subset X$ with U not open in X. Let $Y = \{0, 1\}$. Let the open sets in Y be ϕ, Y, and $\{0\}$.

Define $g : X \longrightarrow Y$, $g(x) = \begin{cases} 0 & \text{for } x \in U \\ 1 & \text{for } x \in X - U \end{cases}$

$\{0\}$ is open in Y and $g^{-1}(\{0\}) = U$, which is not open in X. So g is not continuous. But every function from X to any space Y is continuous. So our assumption that $\exists U \subset X$ with U not open in X has led to a contradiction. So we conclude that every subset of X is open, and thus X has the discrete topology. $\square$

2.10 Problem

Let (Y, T) be a space. Show Y has the indiscrete topology $\iff \forall$ spaces X every function $f : X \longrightarrow Y$ is continuous.

2.11 Result

Let $X = A \bigcup B$, with A and B closed in X. Let $f : A \longrightarrow Y$, and $g : B \longrightarrow Y$ be continuous with $f(x) = g(x) \forall x \in A \bigcap B$.

Define $h: X \longrightarrow Y$, $h(x) = \begin{cases} f(x) & \text{for } x \in A \\ g(x) & \text{for } x \in B \end{cases}$

Claim: h is continuous.

Proof. h is well-defined:
If $x \in A \cap B$, $h(x) = f(x)$ and $h(x) = g(x)$. But by assumption, $f(x) = g(x)$. If follows that h is well-defined.

h is continuous:
Let $C^{closed} \subset Y$. Suppose we knew that $f^{-1}(C) \bigcup g^{-1}(C) = h^{-1}(C)$. Since f, g continuous, $f^{-1}(C)$ and $g^{-1}(C)$ would be closed, by 2.7. By 6.1 on page 69 we would then have $h^{-1}(C)$ is closed. And then, applying 2.7 again (in the opposite direction), we would have h is continuous.

Lemma: $f^{-1}(C) \bigcup g^{-1}(C) = h^{-1}(C)$

Proof of Lemma. Let $x \in f^{-1}(C) \bigcup g^{-1}(C)$.
Then $x \in f^{-1}(C)$ or $x \in g^{-1}(C)$.

Case 1: $x \in f^{-1}(C)$
Then $f(x) \in C$, $x \in A$. So $f(x) = h(x)$. So $h(x) \in C$. That is, $x \in h^{-1}(C)$.
Case 2: $x \in g^{-1}(C)$
Then $g(x) \in C$, $x \in B$. So $h(x) = g(x)$. So $h(x) \in C$. That is $x \in h^{-1}(C)$.
So $f^{-1}(C) \bigcup g^{-1}(C) \subset h^{-1}(C)$

Let $x \in h^{-1}(C)$. So $h(x) \in C$. Since $x \in X = A \bigcup B$, $x \in A$ or $x \in B$, and we have $h(x) = f(x)$ or $h(x) = g(x)$. So $x \in f^{-1}(C)$ or $x \in g^{-1}(C)$.

□

As discussed above, we are done. □

Part III: Homeomorphisms   85

2.12 Result

Let $A \subset X$. Let $f : X \longrightarrow Y$ be continuous.
Let $g : A \longrightarrow Y$, $g(x) = f(x) \, \forall \, x \in A$.
(We call g the **restriction** of f to the subspace A)

Claim: g is continuous.

Proof. Let $U^{open} \subset Y$. Then $f^{-1}(U)$ is open in X, since f is continuous. It would be nice if $g^{-1}(U) = A \cap f^{-1}(U)$. Since $A \cap f^{-1}(U)$ is open in the subspace A, we would have $g^{-1}(U)^{open} \subset A$. And we would be done.

Lemma: $g^{-1}(U) = A \cap f^{-1}(U)$.

Proof of Lemma. Let $x \in g^{-1}(U)$. Then $g(x) \in U$. And since $g(x) = f(x)$, $f(x) \in U$. So $x \in f^{-1}(U)$. And $x \in A$ (since $g(x)$ is defined). So $x \in A \cap f^{-1}(U)$.

Let $x \in A \cap f^{-1}(U)$. Then $x \in A$ and $x \in f^{-1}(U)$. So $f(x) \in U$. But $f(x) = g(x)$. (Note that we know $g(x)$ is defined, since $x \in A$.) So $g(x) \in U$. So $x \in g^{-1}(U)$. □

As discussed above, we are done. □

2.13 Problem

Equip X and Y with the cofinite topology. Find a continuous $f : X \longrightarrow Y$ and a not continuous $g : X \longrightarrow Y$.

2.14 Problem

Let X be a space, $A \subset X$, A have the subspace topology, $f : A \longrightarrow X$, $f(x) = x \, \forall \, x \in A$. We call f the **inclusion** of

A into X. Show f is continuous.
(Hint: Let $U^{open} \subset X$. Show $f^{-1}(U) = U \bigcap A$.)

2.15 Problem

Let X, Y be spaces. Let $t \in Y$. Let $f : X \longrightarrow Y$, $f(x) = t \,\forall\, x \in X$. (That is, let f be the constant function that sends all elements of X to t.) Show f is continuous.

Chapter 3

Homeomorphisms

3.1 Definition

Let $f : A \longrightarrow B$. f is said to be a **homeomorphism** between A and B if f is continuous, f^{-1} is continuous and f is bijective. If such an f exists, we say A and B are **homeomorphic**.

3.2 Result

Let (X, T) be a space.

Claim: (X, T) is homeomorphic to itself.

Proof. $id : X \longrightarrow X$ is continuous by 2.2 on page 79.
$id^{-1} = id$, so id^{-1} is continuous. And the identity is a bijection. So we have constructed a homeomorphism between X and itself. So (X, T) is homeomorphic to itself. $\square$

3.3 Result

Let $X = (0,1)$. Let the open intervals in $\mathbb{R}$ contained in $(0,1)$ be a basis for a topology T on X.

Let $Y = (a,b)$ where $a, b \in \mathbb{R}$ and $a < b$. Let the open intervals in $\mathbb{R}$ contained in (a,b) be a basis for a topology T' on Y.

Claim: (X, T) is homeomorphic to (Y, T')

Proof. Let $f : X \longrightarrow Y$, $f(x) = (1-x)a + x(b)$
f is well-defined.

f is 1-1:
Suppose that $f(x) = f(y)$, for some $x \in X$, $y \in Y$.
Then $(1-x)a + x(b) = (1-y)a + y(b)$.
So $((1-x) - (1-y))a = (y-x)b$
So $(y-x)a = (y-x)b$.
But a does not equal b. So $y - x = 0$. So $x = y$.

f is onto:
Let $y \in Y$. Let $t = \frac{y-a}{b-a}$ $y = f(t)$ (check this). We need to check though that t is defined and $t \in X$.
t is defined since $a \neq b$.
$y \in Y$, $y < b$. So $t < 1$.
And since $y > a$ and $b > a$, $t > 0$. So $t \in X$. So f is onto.

f is continuous:
Let (c,d) be a basis element for T'. $f^{-1}(c,d) = (\frac{c-a}{b-a}, \frac{d-a}{b-a})$, which is open in X. (Why?) So f is continuous.

f^{-1} continuous:
Let (c,d) be a basis element for topology on X.
Then $f((c,d)) = ((1-c)a + bc, (1-d)a + bd)$ which is open in Y (Why?).

So f^{-1} is continuous.

□

3.4 Result

Claim: Homeomorphism is an equivalence relation.

Proof. 1) X is homeomorphic to X by 3.2

2) Suppose X is homeomorphic to Y. Then $\exists$ a homeomorphism $f : X \longrightarrow Y$. We need a homeomorphism $g : Y \longrightarrow X$. Let's try f^{-1}.

Lemma 1: $f^{-1} : Y \longrightarrow X$ is a homeomorphism.

Proof of Lemma 1. a) f^{-1} is well-defined:
Let $y \in Y$. $\exists$ a unique $x \in X$ such that $f(x) = y$, since f is bijective. $f^{-1}(y) = x$. So f^{-1} is well-defined.

b) f^{-1} is 1-1:
Suppose $f^{-1}(x) = f^{-1}(y)$. Then, since f is well-defined, $x = y$.

c) f^{-1} is onto:
Let $x \in X$. Then, since f is well-defined, $\exists y \in Y$ such that $f(x) = y$. $f^{-1}(y) = f^{-1}(f(x)) = x$. So f^{-1} is onto.

d) $(f^{-1})^{-1} = f$ is continuous since f is a homeomorphism.

e) f^{-1} is continuous, since f is a homeomorphism.

So f^{-1} is a homeomorphism.

□

So Y is homeomorphic to X and the homeomorphism relation is symmetric.

3) Suppose X is homeomorphic to Y and Y is homeomorphic to Z. Then there are homeomorphisms $f : X \longrightarrow Y$, and $g : Y \longrightarrow Z$. We want to find a homeomorphism from X to Z. Let's try $g \circ f$.

Lemma 2: Let $h = g \circ f$. $h : X \longrightarrow Z$ is a homeomorphism.

Proof of Lemma 2. a) h is well-defined:
$f(x) = y$ for a unique $y \in Y$ and $g(y) = z$ for a unique $z \in Z$. $h(x) = g(f(x)) = g(y) = z$. So h is well-defined.

b) h is onto:
Let $z \in Z$. $\exists\, y \in Y$ such that $g(y) = z$, since g is onto. $\exists\, x \in X$ such that $f(x) = y$, since f is onto. And $h(x) = g(f(x)) = z$. So h is onto.

c) h is 1-1:
Suppose $h(a) = h(b)$ for some $a, b \in X$. Then $g(f(a)) = g(f(b))$. Since g is 1-1, $f(a) = f(b)$. Since f is 1-1, $a = b$. So h is 1-1.

d) h is continuous:
2.8 on page 82

e) h^{-1} is continuous: 3.24 on page 35 and 2.8 on page 82 □

So h is a homeomorphism and the homeomorphism relation is transitive.

□

Part III: Homeomorphisms

3.5 Problem

Let $\{0,1\}$ and $\{0,1,2\}$ be equipped with some topologies. Show there is no homeomorphism $h : \{0,1\} \longrightarrow \{0,1,2\}$.

3.6 Problem

Let A, B be finite sets. Let A have n elements, and B have m elements, where $n \neq m$. Equip A and B with any topologies. Show A is not homeomorphic to B.

3.7 Problem

Let C be a finite set, and let D be an infinite set. Equip C and D with any topologies. Show C is not homeomorphic to D.

3.8 Problem

Let E, F be sets. Suppose there does not exist bijective $f : E \longrightarrow F$. Equip E and F with any topologies. Show E is not homeomorphic to F.

3.9 Definition

Suppose a space can have a certain property.

We say that a property is a **topological property** when homeomorphic spaces must either both have the property or both not have the property.

Suppose we are interested in showing that a space X is not homeomorphic with Y. One approach is to show that there is no function from X to Y that is a homeomorphism. Sometimes this is easy. Sometimes it is hard.

Another approach is to show that there is some topological property that X has and Y does not have. This is sometimes easier.

Some topological properties will be introduced in the next part.

Part IV

More Properties

Chapter 1

Retractions

1.1 Definition

Let $A \subset X$. Let $f : X \longrightarrow A$. We call f a **retraction** of X onto A if f is continuous and $f(a) = a \,\forall\, a \in A$. A is said to be a **retract** of X if such an f exists.

1.2 Result

Let X be a space. Let A be non-empty, $A \subset X$. Equip A with the discrete topology.

Claim: A is a retract of X.

Proof. Choose any $a_1 \in A$.
Define $r : X \longrightarrow A$, $r(x) = \begin{cases} x & \text{for } x \in A \\ a_1 & \text{for } x \in X - A \end{cases}$
By 2.11 on page 83 r is continuous. So we're done. $\square$

1.3 Result

Let A be a retract of X, and $h : X \longrightarrow Y$ be a homeomorphism.

Claim: $h(A)$ is a retract of Y.

Proof. Since A is a retract of X, we have a continuous $r : X \longrightarrow A$, $r(a) = a \; \forall \, a \in A$. To show $h(a)$ is a retract of Y, we need to construct a retraction of Y onto $h(A)$.

Let $f = h \circ r \circ h^{-1}$. $f : Y \longrightarrow h(A)$. f is the composition of continuous functions, thus continuous.

Let $q \in h(A)$. Then $q = h(t)$ for some $t \in A$.
$f(q) = f(h(t)) = h(r(h^{-1}(h(t)))) = h(r(t))$. Since r is a retraction of X onto A and $t \in A$, we have $r(t) = t$. So $f(q) = f(h(t)) = h(t) = q$.
So $h(A)$ is a retract of Y, with retraction f. □

1.4 Result

Let $B \subset A \subset X$.
Let $f : X \longrightarrow A$ and $g : A \longrightarrow B$ be retractions. Let $h : X \longrightarrow B$, $h = g \circ f$.

Claim: h is a retraction of X onto B.

Proof. Since f and g are continuous, we know h is continuous. Let $b \in B \subset A$. Then $h(b) = g(f(b))$. Since $b \in A$, we have $f(b) = b$. So $h(b) = g(b)$. But, since $b \in B$, $g(b) = b$. So $h(b) = b$.
So h is a retraction of X onto B, and B is a retract of X. □

Part IV: More Properties   97

1.5 Problem

Let $A \subset X$, let X be a space. Let $f : X \longrightarrow A$ be a retraction. Show f is onto.

1.6 Problem

Let $(0,1) \subset \mathbb{R}$ be equipped with any topology. Let $\{0\}$ have the discrete topology. Show $\{0\}$ is not a retract of $(0,1)$.

What happens if we only require $\{0\}$ be equipped with some topology (instead of specifically the discrete topology)?

Chapter 2

Fixed Point Property

2.1 Definition

Let $f : A \longrightarrow A$. When $f(a) = a$, we say a is a **fixed point** of f.

2.2 Definition

Let X be a space. X is said to have the **fixed-point property**, when every continuous function $f : X \longrightarrow X$ has a fixed point. This will also sometimes be worded as 'X is FPP'.

2.3 Result

Claim: The fixed point property is a topological property.

Proof. Let X be FPP. Let $h : X \longrightarrow Y$ be a homeomorphism. We want to show that Y is FPP. So, let $g : Y \longrightarrow Y$ be continuous. We want to show that g has a fixed point. Let $j = h^{-1} \circ g \circ h$. $j : X \longrightarrow X$. j is continuous by 2.8 on page 82. Since X is FPP, $\exists\, t \in X$ such that $j(t) = t$.

$j(t) = h^{-1}(g(h(t)))$. So $h(j(t)) = h(h^{-1}(g(h(t))) = g(h(t))$. But $j(t) = t$, so we have $h(t) = g(h(t))$. So $h(t)$ is a fixed point of g. So Y is FPP. □

2.4 Example

Let $X = \{x\}$. Equip X with the discrete topology.

Claim: X is FPP.

Proof. Let $f : X \longrightarrow X$ be continuous. Since f is well-defined, $f(x) = x$. (Note that there is precisely one function from X to X, and that function is continuous.) □

2.5 Example

Let $a, b \in X$, $a \neq b$, $X = \{a, b\}$. Let X have the discrete topology.

Claim: X is not FPP.

Proof. Let $f : X \longrightarrow X, f(a) = b, f(b) = a$. f is continuous and has no fixed point. □

2.6 Result

Claim: $\mathbb{R}$ is not FPP.

Proof. Let $f : \mathbb{R} \longrightarrow \mathbb{R}$, $f(x) = x + 1$. f is continuous and has no fixed point. □

2.7 Result

Let $r : X \longrightarrow A$ be a retraction, with X FPP.

Claim: A is FPP.

Proof. Let $f : A \longrightarrow A$ be continuous.
Let $h : A \longrightarrow X$, $h(a) = a \; \forall \, a \in A$.
Let $t = h \circ f \circ r$. h is continuous by 2.14 on page 86.
So $t : X \longrightarrow X$ is continuous by 2.8 on page 82.
Since X is FPP, t has a fixed point, x. $t(x) = h(f(r(x))) = x$.
But $h(f(r(x))) = f(r(x))$ (why?).

It would be nice if x were in A. Then $r(x) = x$, and $f(r(x)) = f(x) = x$.
And we will have shown that A is FPP.

Lemma: $x \in A$.

Proof of Lemma. We know $t(x) \in A$, since $h(A) \subset A$. But $t(x) = x$. So $x \in A$.

□

□

2.8 Remark

We still haven't found a non-trivial space that has the fixed-point property. In the next chapter, we will introduce the topological property of connectedness. Then we will use this idea to find some non-trivial spaces that have the fixed point property.

Chapter 3

Connectedness

3.1 Definition

Let $A, B \subset X$. Then A and B are said to be **disjoint** when $A \bigcap B = \phi$.

3.2 Definition

Let X be a space and let $A, B \subset X$. A and B form a **separation** of X when $X = A \bigcup B$, and A and B are open, disjoint, and non-empty.

3.3 Definition

Let X be a space. Then X is **connected** when there is no separation of X.

3.4 Result

Let X be a space.

Claim: X is connected $\iff$ the only sets in X that are both open and closed are ϕ and X.

Proof. $\implies$ Let X be connected. Let $U \subset X$, and let U be open and closed. Then $X = U \bigcup (X - U)$. Since U is closed, $X - U$ is open. We have $X = U \bigcup (X - U)$. Both U and $X - U$ are open. And U and $X - U$ are disjoint. Since X is connected, either U or $X - U$ must be empty. So $U = \phi$ or $U = X$.

$\impliedby$ Let the only sets in X that are both open and closed be ϕ and X. Suppose U and V form a separation of X. Then $U = X - V$, which is open. So U is closed (and open). So $U = \phi$ or X.

Case 1: $U = \phi$
This cannot happen, since U is non-empty, by definition of a separation.

Case 2: $U = X$
Then $V = \phi$, which can also not happen, by definition of a separation.

So our assumption that there is a separation of X is false. So X is connnected. $\square$

3.5 Example

Let $X = (0,1) \bigcup [2,3] \subset \mathbb{R}$. Equip X with the subspace topology (that is, the topology in 4.12 on page 62).

Claim: X is not connected.

Proof. By 4.12 on page 62 and 3.4 $\square$

Part IV: More Properties   105

3.6 Result

Let $f : X \longrightarrow Y$ be continuous. Let $U^{connected} \subset X$.

Claim: $f(U)^{connected} \subset Y$.

Proof. Suppose that $f(U)$ is not connected. Then $\exists\, A, B$ that form a separation of $f(U)$. It would be nice if $f^{-1}(A)$ and $f^{-1}(B)$ formed a separation of U. That would contradict the fact that U is connected. That would mean our assumption that $f(U)$ is not connected was false. That would tell us $f(U)^{connected} \subset Y$.

Lemma: $f^{-1}(A)$ and $f^{-1}(B)$ form a separation of U.

Proof of Lemma. non-empty:
Since A and B are non-empty and subsets of $f(U)$, $f^{-1}(A)$ and $f^{-1}(B)$ are non-empty.

disjoint:
Suppose $x \in f^{-1}(A) \bigcap f^{-1}(B)$.
Then $f(x) \in A$ and $f(x) \in B$. But A and B are disjoint. So our assumption that there is an $x \in f^{-1}(A) \bigcap f^{-1}(B)$ is false. So $f^{-1}(A) \bigcap f^{-1}(B) = \phi$.

open:
A and B are open in Y (since they form a separation of U). Since f is continuous, $f^{-1}(A)$ and $f^{-1}(B)$ are open in X.
$\square$

So $f(U)^{connected} \subset Y$ and we are done. $\square$

3.7 Result

Claim: Connectedness is a topological property.

Proof. Let X be homeomorphic to Y by a homeomorphism f. Let X be connnected. We want to show that Y is connected.

Suppose that Y is not connected. Then $\exists\, U \subset Y$ such that $U \neq \phi$, $U \neq Y$, and U is open and closed in Y. $f^{-1}(U)$ is open and closed in X (since f is continuous), non-empty (since f is onto and $U \neq \phi$), and not X (since f is onto and $U \neq Y$). So X is not connected. But X is connected. So our assumption that Y is not connected has led to a contradiction. So Y is connected. □

3.8 Remark

Equip $\mathbb{R}$ with the standard topology. It is connected. Let A be any interval in $\mathbb{R}$, and equip it with the subspace topology. Then A is connected. We do not prove either of these results, but we will use them. Many topology books contain proofs of these results, including [9].

3.9 Result

Let $\mathbb{Q} \subset \mathbb{R}$ have the subspace topology.

Claim: $\mathbb{Q}$ is not connected.

Proof. $\mathbb{Q} \cap (-\infty, \sqrt{2})$ is open in $\mathbb{Q}$. $\mathbb{Q} \cap (\sqrt{2}, \infty)$ is open in $\mathbb{Q}$. Let $U = \mathbb{Q} \cap (-\infty, \sqrt{2})$ and let $V = \mathbb{Q} \cap (\sqrt{2}, \infty)$. $U \cup V = \mathbb{Q}$, by the Lemma in 3.15 on page 31. U and V are disjoint. U and V are open. So we have found a separation of $\mathbb{Q}$. So $\mathbb{Q}$ is not connected. □

3.10 Result

Let X be a space, let $\{0,1\}$ have the discrete topology.

Claim: X is connected $\iff f : X \longrightarrow \{0,1\}$ being continuous implies f is constant.

Proof. $\implies$ Let X be connected. Let $f : X \longrightarrow \{0,1\}$ be continuous. We want to show that f must be constant. Suppose f is not constant. Then $\exists\, x, y \in X$ such that $f(x) = 0, f(y) = 1$. $f^{-1}(\{0\}) \neq \phi, f^{-1}(\{1\}) \neq \phi$. Since f is continuous, $f^{-1}(\{0\})$ and $f^{-1}(\{1\})$ are open in X. They are also disjoint, since f is a function. Also, $X = f^{-1}(\{0\}) \bigcup f^{-1}(\{1\})$. So $f^{-1}(\{0\}) \bigcup f^{-1}\{1\}))$ forms a separation of X. But X is connected. So our assumption that f is not constant has led to a contradiction. So f is constant.

$\impliedby$ Suppose every continuous $f : X \longrightarrow \{0,1\}$ is constant. We want to show that X is connected. Suppose not. Then $X = U \bigcup V$, a separation.
Define $f : X \longrightarrow \{0,1\}, f(x) = \begin{cases} 0 & \text{for } x \in U \\ 1 & \text{for } x \in V \end{cases}$
$f^{-1}(\{0\}) = U$ which is open. $f^{-1}(\{1\}) = V$ which is open. So f is continuous. So f is constant. But f is not constant. So our assumption that X is not connected has led to a contradiction. So X is connected. $\square$

3.11 Result

Let X be FPP.

Claim: X is connected.

Proof. Suppose not. Then $X = U \bigcup V$, a separation. U and V are closed and disjoint. $\exists\, u \in U, v \in V$.

Define $f: X \longrightarrow X, f(x) = \begin{cases} v & \text{for } x \in U \\ u & \text{for } x \in V \end{cases}$

By 2.11 on page 83 f is continuous. That's bad, because f has no fixed point. And X is FPP. So our assumption that X is not connected was false. So X is connected. □

3.12 Definition

Let X be a space. Let $x \in X$. We call the union of all connected subsets of X that contain x the **component** of x in X.

3.13 Problem

Show ϕ is connected.

3.14 Problem

Let X be a space, let $x \in X$, and give $\{x\} \subset X$ the subspace topology. Show $\{x\}$ is connected.

3.15 Problem

Let X have the indiscrete topology. Show X is connected.

3.16 Problem

Let $X = \{0, 1, \frac{1}{2}, \frac{1}{3}, \ldots\}$. Give X the topology it inherits as a subspace of $\mathbb{R}$. Show X is not connected.

3.17 Problem

Let X have the discrete topology.
Show X is connected $\iff X = \phi$ or X has precisely one element.

3.18 Problem

Equip a finite set X with the cofinite topology. Show X is connected.

3.19 Problem

Let U be a finite subset of $\mathbb{R}$. Equip $\mathbb{R}$ with its standard topology and U with the subspace topology. Show U is not connected.

3.20 Remark

Let $a, b \in \mathbb{R}, a < b$. Let $f : [a, b] \longrightarrow \mathbb{R}$ be continuous. Then f takes on every value between $f(a)$ and $f(b)$. That is, if $f(a) < f(b)$, then for every y where $f(a) < y < f(b)$ $\exists\, x \in [a, b]$ so that $f(x) = y$. And if $f(b) < f(a)$, then for every y where $f(a) > y > f(b)$ $\exists\, x \in [a, b]$ so that $f(x) = y$. Consider the case where $f(a) = f(b)$... why is the condition trivial in that case? The result just described is a version of the Intermediate Value Theorem. It is true, and we will not prove it. See [9] for a proof.

3.21 Result

Suppose you know that $[0, 1] \subset \mathbb{R}$ is connected, and you know the Intermediate Value Theorem is true (both are true, but we have proved neither).

Claim: $[0, 1]$ is FPP.

Proof. Let $f : [0, 1] \longrightarrow [0, 1]$ be continuous. We want to show f has a fixed point.

Case 1: $f(0) = 0$ or $f(1) = 1$
Then f has a fixed point.

Case 2: $f(0) \neq 0$ and $f(1) \neq 1$
Then $f(0) > 0$ and $f(1) < 1$. Define $g : [0, 1] \longrightarrow \mathbb{R}$, $g(x) = x - f(x)$.
g is continuous. Note that $g(0) = 0 - f(0) < 0$ and $g(1) = 1 - f(1) > 0$. By the Intermediate Value Theorem, $\exists\, t \in [0, 1]$ such that $g(t) = 0$. But $g(t) = t - f(t)$. So we have a t so that $t = f(t)$, and that t is a fixed point of f.
So, given an arbitrary continuous $f : [0, 1] \longrightarrow [0, 1]$, we have shown that f has a fixed point. So $[0, 1]$ has the fixed point property. $\square$

Part V

More Examples

Chapter 1

Subspace Topology

1.1 Result

Let (X, T') be a topological space and let $A \subset X$.
Let $T = \{U;\ U = V \cap A \text{ for some } V \in T'\}$

Claim: (A, T) is a space.

Proof. 1) $\phi \in T'$. $\phi \cap A = \phi$. So $\phi \in T$.
$X \in T'$. $X \cap A = A$. So $A \in T$.

2) Suppose you have a collection $\{U_\alpha\}_{\alpha \in J}$, $U_\alpha^{open} \subset T \forall \alpha \in J$.
Each $U_\alpha = V_\alpha \cap A$ for some $V_\alpha \in T'$.
$\bigcup_{\alpha \in J} U_\alpha = \bigcup_{\alpha \in J}(V_\alpha \cap A) = (\bigcup_{\alpha \in J} V_\alpha) \cap A$. (Why?)
Since each V_α is open in X, $\bigcup_{\alpha \in J} V_\alpha$ is open in X.
So $\bigcup_{\alpha \in J} U_\alpha$ is open in A.

3) Let $U_1, U_2 \in T$. Then $U_1 = V_1 \cap A$ for some $V_1 \in T'$.
And $U_2 = V_2 \cap A$ for some $V_2 \in T'$.
So $U_1 \cap U_2 = (V_1 \cap A) \cap (V_2 \cap A) = (V_1 \cap V_2) \cap A$.
Since T' is a topology, $V_1 \cap V_2 \in T'$.
So $U_1 \cap U_2$ is open in A. □

1.2 Definition

Let (X, T') be a space. Let $A \subset T$, and let T be defined as in 1.1. Then we call T the **subspace topology** on A. We say A inherits this topology as a subset of X.

1.3 Example

Let $(0,1) = A \subset \mathbb{R}$. Let $\mathbb{R}$ have the standard topology and give A the subspace topology. Let $(0, \frac{1}{2}] = B$.

Claim: B is closed in A.

Proof. We want to show that $A - B$ is open in A.
$A - B = (\frac{1}{2}, 1)$.

To show this is open in A we need to show that $(\frac{1}{2}, 1) = U \cap A$ for some $U^{open} \subset \mathbb{R}$. Let $U = (\frac{1}{2}, 1)$. $U^{open} \subset \mathbb{R}$, and $U \cap A = (\frac{1}{2}, 1)$.

So $A - B$ is open in A. So B is closed in A. $\square$

1.4 Problem

Let $(0,1) \subset \mathbb{R}$. Let $\mathbb{R}$ have the standard topology, and let $(0,1)$ have the subspace topology. Show $[\frac{1}{4}, \frac{3}{8}]$ is closed in $(0,1)$.

Proof. $(0,1) - [\frac{1}{4}, \frac{3}{8}] = (0, \frac{1}{4}) \bigcup (\frac{3}{8}, 1)$.
Use this to show $[\frac{1}{4}, \frac{3}{8}]$ is closed in $(0,1)$. $\square$

1.5 Remark

Note that 1.3 and 1.4 tell us that a closed set in the subspace topology may or may not be closed in the big space.

1.6 Problem

Let $A = [0,1]$. Consider A as a subspace of $\mathbb{R}$ with the standard topology. Show $(0, \frac{1}{2}]$ is not closed in A.

1.7 Result

Let $\mathbb{R}$ have the standard topology, Let T be the subspace topology $\mathbb{N}$ inherits from $\mathbb{R}$. Let T' be the discrete topology on $\mathbb{N}$.

Claim: $T = T'$

Proof. $T \subset T'$:
In the discrete topology on $\mathbb{N}$, every subset of $\mathbb{N}$ is open. So if $U \in T$ then $U \subset \mathbb{N}$ and thus $U \in T'$.

$T' \subset T$:
Let $U \in T'$. Suppose we knew that $\{x\} \in T \ \forall \ x \in \mathbb{N}$. Then we would know that U would have to be open (since $U = \bigcup_{x \in U} \{x\}$). Let $x \in \mathbb{N}$. Then $\{x\} = (x - \frac{1}{2}, x + \frac{1}{2}) \bigcap \mathbb{N}$. So $\{x\} \in T$. By the above comment, $T' \subset T$ and we are done. □

1.8 Problem

Let $\mathbb{R}$ have the standard topology, $\mathbb{Z} \subset \mathbb{R}$. Give $\mathbb{Z}$ the subspace topology. What familiar topology is this? (Hint: it is similar to 1.7)

1.9 Problem

Let X have the discrete topology. Let $A \subset X$ have the subspace topology, T. Show T is the same as the discrete topology on A.

1.10 Problem

Let X have the indiscrete topology. Let $A \subset X$ have the subspace topology, T. Show T is the same as the indiscrete topology on A.

1.11 Problem

Let $\mathbb{R}$ have the standard topology, $[0,1] \subset \mathbb{R}$ the subspace topology. Let $a, b \in \mathbb{R}$, $0 < a < b < 1$. Show $[0, a)$ and $(b, 1]$ are open in $[0, 1]$.

1.12 Result

Let $\mathbb{R}$ have the standard topology. Let $[0, 1] \subset \mathbb{R}$ have the subspace topology.

Claim: $[0, 1]$ is a retract of $\mathbb{R}$.

Proof. Define $f : \mathbb{R} \longrightarrow [0, 1]$, $f(x) = \begin{cases} 0 & \text{for } x < 0 \\ 1 & \text{for } x > 1 \\ x & \text{for } x \in [0, 1] \end{cases}$

f is well-defined.

We need to show f is continuous.
Let $B = \{U \subset [0,1]; U = [0,a) \text{ or } U = (b,1] \text{ or } U = (a,b)\}$ for $a, b \in \mathbb{R}$ and $0 < a < b < 1$. B is a basis for the topology

on $[0,1]$ (Why?). Let $B_1 \in B$. Then $f^{-1}(B_1) = (-\infty, a)$ or $f^{-1}(B_1) = (a, \infty)$ or $f^{-1}(B_1) = (a, b)$. So $f^{-1}(B_1)$ is open in $\mathbb{R}$ for every basis element B_1. So f is continuous. So $[0,1]$ is a retract of $\mathbb{R}$. □

1.13 Problem

Let $\mathbb{R}$ have the standard topology, and let $[0,1) \subset \mathbb{R}$ have the subspace topology. Show $[0,1)$ is not FPP.

1.14 Problem

1) Is the subspace of an FPP space also FPP?
2) Examine subspaces of retracts.
3) Examine subspaces of a space with the cofinite topology.

1.15 Problem

Let X be a space, let $A \subset X$ have the subspace topology. Let $C \subset A$. Show $C^{closed} \subset A \implies \exists\, V^{closed} \subset X$ such that $C = V \bigcap A$.

1.16 Remark

Consider $[0,1]$ as a subspace of $\mathbb{R}$ with the standard topology. Try to find a closed set in $[0,1]$ that is not closed in $\mathbb{R}$. (Hint: Dont try for too long.)

1.17 Problem

Let X be a space. Suppose $A \subset B \subset X$, and B is equipped with the subspace topology. Suppose A is closed in B and B is closed in X. Show A is closed in X. (Hint: 1.15 is useful)

1.18 Remark

Why does this explain the hint in 1.16?

Chapter 2

Lower Limit Topology, K-Topology

2.1 Result

Let $B = \{[a,b); a, b \in \mathbb{R}, \text{ and } a < b\}$

Claim: B is a basis for a topology on $\mathbb{R}$.

Proof. 1) Let $x \in \mathbb{R}$. $x \in [x, x+1) \in B$

2) Let $x \in B_1, B_2$. Then $B_1 = [a_1, b_1)$ and $B_2 = [a_2, b_2)$ for some a_1, a_2, b_1, b_2 $\in \mathbb{R}$ with $a_1 < b_1$ and $a_2 < b_2$. Let $a = max\{a_1, a_2\}$. Let $b = min\{b_1, b_2\}$. Let $B_3 = [a, b)$. $B_3 \in B$, and $x \in B_3 \subset B_1 \bigcap B_2$. □

2.2 Definition

When we equip $\mathbb{R}$ with the the topology generated by the basis $\mathbb{B}$ from 2.1, $\mathbb{R}$ is said to be equipped with the **lower limit topology**.

2.3 Result

Let $K = \{\frac{1}{n}; n \in \mathbb{N}\}$.
Let $B = \{U; U = (a,b) \text{ or } U = (a,b) - K \text{ for some } a, b \in \mathbb{R} \text{ with } a < b\}$

Claim: B is a basis for a topology on $\mathbb{R}$.

Proof. 1) Let $x \in \mathbb{R}$. Then $x \in (x-1, x+1) \in B$.

2) Let $x \in B_1 \cap B_2$, $B_1, B_2 \in B$.

Let's call basis sets of the form (a,b) Type 1, and let's call basis sets of the form $(a,b) - K$ Type 2.

Case 1: B_1, B_2 are both Type 1.

Then $B_1 = (a_1, b_1)$, $B_2 = (a_2, b_2)$. Let $a = max\{a_1, a_2\}$ and let $b = min\{b_1, b_2\}$. Let $B_3 = (a, b)$. Then $B_3 \in B$, and $x \in B_3 \subset B_1 \cap B_2$.

Case 2: B_1, B_2 are both Type 2.

Then $B_1 = (a_1, b_1) - K$, $B_2 = (a_2, b_2) - K$. Let $a = max\{a_1, a_2\}$ and let $b = min\{b_1, b_2\}$. Let $B_3 = (a,b) - K$. Then $B_3 \in B$, and $x \in B_3 \subset B_1 \cap B_2$.

Case 3: B_1 is Type 1, and B_2 is Type 2.

Then $B_1 = (a_1, b_1)$, $B_2 = (a_2, b_2) - K$. Let $a = max\{a_1, a_2\}$ and let $b = min\{b_1, b_2\}$. Let $B_3 = (a,b) - K$. Then $B_3 \in B$, and $x \in B_3 \subset B_1 \cap B_2$. □

Part V: More Examples   121

2.4 Definition

When we equip $\mathbb{R}$ with the topology generated by the basis B in 2.3, we say $\mathbb{R}$ is equipped with the **K-topology**.

2.5 Problem

Equip $\mathbb{R}$ with the lower limit topology. Let $a, b \in \mathbb{R}$, $a < b$. Show (a, b) is open in $\mathbb{R}$.
(Hint: Show that $(a, b) = \bigcup_{x \in (a,b)} [x, b)$)

2.6 Problem

Equip $\mathbb{R}$ with the lower limit topology. Let $a \in \mathbb{R}$. Show $(-\infty, a)$ is both open and closed in $\mathbb{R}$.

2.7 Problem

Equip $\mathbb{R}$ with the lower limit topology. Let $a \in \mathbb{R}$. Show $[a, \infty)$ is both open and closed in $\mathbb{R}$.

2.8 Problem

Equip $\mathbb{R}$ with the lower limit topology. Let $a \in \mathbb{R}$. Show (a, ∞) is both open and closed in $\mathbb{R}$.

2.9 Problem

Let T be the lower limit topology on $\mathbb{R}$. Show $(\mathbb{R}, T)$ is not connected.

2.10 Problem

Equip $\mathbb{R}$ with the lower limit topology. Let $a, b \in \mathbb{R}, a < b$. Show $[a, b]$ is not open.

2.11 Result

Let T be the standard topology on $\mathbb{R}$, and let T' be the K topology.

Claim: $T \subset T'$

Proof. Let $U \in T$. Recall that a basis for T is the set of all open intervals in $\mathbb{R}$. So U is a union of open intervals in $\mathbb{R}$. So, by 2.4, $U \in T'$ (Why?). $\square$

Chapter 3

Sierpinski Space

3.1 Definition

Let $X = \{0, 1\}$. Let $T = \{\phi, \{0, 1\}, \{0\}\}$. We call (X, T) the **Sierpinski space**.

3.2 Example

Let X be the Sierpinski space. Let Y be a set with precisely two elements x and y. Let the open sets in Y be ϕ, Y, and $\{x\}$.

Claim: X is homeomorphic to Y.

Proof. To show X is homeomorphic to Y, we will construct a homeomorphism $f : X \longrightarrow Y$. Let $f(0) = x$, and $f(1) = y$. We show that this defines our desired homeomorphism.

f is well-defined and f is bijective.
f is continuous:
There are three open sets in Y. We'll take the inverse image of each and make sure it is open.

$f^{-1}(\phi) = \phi$, which is open in X.
$f^{-1}(Y) = X$, which is open in X.
$f^{-1}(\{x\}) = \{0\}$, which is open in X.
So f is continuous.

f^{-1} is continuous:
Since $(f^{-1})^{-1} = f$, we will just check that the image of each open set in X under f is open in Y.

$f(\phi) = \phi$ which is open in Y.
$f(X) = Y$ which is open in Y.
$f(\{0\}) = \{x\}$ which is open in Y.
So f^{-1} is continuous.

So we are done. $\square$

3.3 Result

Claim: The Sierpinski space is connected.

Proof. Let X be the Sierpinski space. Suppose X is not connected. Then $\exists\, U, V$ that form a separation of X. Since both U and V are open, non-empty, and disjoint, U or V has to equal $\{0\}$. (Why?) Without loss of generality, say $U = \{0\}$. Since $X = U \bigcup V$ and U and V are disjoint, V must equal $\{1\}$. And V is open, by assumption. But $V = \{1\}$ is not open. So our assumption that X is not connected has led to a contradiction. So X is connected. $\square$

3.4 Problem

Let Y be the Sierpinski space. Both Y and $\mathbb{R}$ are connected. Note that this does not tell us that Y is homeomorphic to $\mathbb{R}$. Their both being connected is not helpful in deciding whether they are homeomorphic or not. We do however know that Y is not homeomorphic to $\mathbb{R}$. How?

3.5 Problem

Let X be the Sierpinski Space. What are all the closed sets of X? What are all the open sets of X?

3.6 Result

Let X be the Sierpinski space.

Claim: X is FPP.

Proof. Let $f : X \longrightarrow X$ be continuous. Since $\{0\}$ is open, $f^{-1}(\{0\})$ must be open.
Case 1: $f^{-1}(\{0\}) = X$
Then $f(0) = 0$, and f has fixed point 0.
Case 2: $f^{-1}(\{0\}) = \phi$
Then $f(1) = f(0) = 1$. And f has fixed point, 1.
Case 3: $f^{-1}(\{0\}) = \{0\}$
Then $f(0) = 0$, and f has fixed point 0. So X is FPP. □

Chapter 4

Path Connected

4.1 Definition

Let X be a space. Let $[0,1] \subset \mathbb{R}$ have the subspace topology. Let $a, b \in X$. We say f is a **path** in X from a to b when $f : [0,1] \longrightarrow X$, f is continuous, $f(0) = a$, and $f(1) = b$.

4.2 Example

Let $f : [0,1] \longrightarrow \mathbb{R}$, $f(x) = (1-x)(a) + x(b)$. Let $a, b \in \mathbb{R}$.

Claim: f is a path from a to b.

Proof. f is continuous. $f(0) = a$. $f(1) = b$. □

4.3 Definition

Let X be a space. We say X is **path connected** if the following is true:
$\forall\, x, y \in X\ \exists$ a path in X from x to y.

4.4 Result

Let $X = \phi, Y = \{y\}$.

Claim: X and Y are both path connected.

Proof. X and Y satisfy path conectedness vacuously. □

4.5 Result

Claim: $\mathbb{R}$ is path connected.

Proof. 4.2 □

4.6 Result

Claim: The Sierpinski Space is path connected.

Proof. $X = \{0,1\}$. Let $T = \{\phi, \{0,1\}, \{0\}\}$. Let $x, y \in X, x \neq y$. Then, without loss of generality, we can say $x = 0$ and $y = 1$. We want to construct a path from x to y. So we want a function $f : [0,1] \longrightarrow X$ that is continuous such that $f(0) = x, f(1) = y$. Let $f(0) = 0, f(1) = 1, f(t) = 0 \; \forall \, t \in (0,1)$. We check that f is continuous:

$f^{-1}(\phi) = \phi$, which is open in $[0,1]$.

$f^{-1}(X) = [0,1]$, which is open in $[0,1]$.

$f^{-1}(\{0\}) = [0,1) = (-\frac{1}{2}, 1) \bigcap X$. Since $(-\frac{1}{2}, 1)$ is open in $\mathbb{R}$, $(-\frac{1}{2}, 1) \bigcap X$ is open in $[0,1]$.

So we have shown that f is continuous.

So we have shown that X (which is the Sierpinski Space) is path connected. □

4.7 Result

Let X be path connected and $f : X \longrightarrow Y$ be a homeomorphism.

Claim: Y is path connected.

Proof. Let $x, y \in Y, x \neq y$. Since f is well-defined, $f^{-1}(x) \neq f^{-1}(y)$. So we have a path g in X from $f^{-1}(x)$ to $f^{-1}(y)$. That is $g : [0,1] \longrightarrow X$, g is continuous, $g(0) = f^{-1}(x)$ and $g(1) = f^{-1}(y)$. Let's look at $f \circ g : [0,1] \longrightarrow Y$. It would be nice if $f \circ g$ was a path from x to y.

Lemma: $f \circ g$ is a path from x to y.

Proof of Lemma: $f \circ g$ is continuous, by 2.8 on page 82.
$(f \circ g)(0) = f(g(0)) = f(f^{-1}(x)) = x$
$(f \circ g)(1) = f(g(1)) = f(f^{-1}(y)) = y$
□

By the Lemma, we are done. □

4.8 Result

Claim: Path connectedness is a topological property

Proof. 4.7 □

4.9 Result

Let X be path connected.

Claim: X is connected.

Proof. Let $x, y \in X$. We show x and y are in the same component of X. If this is true, then we have shown that there is only one component of X. That is, X is connected.

Since X is path connected, we can choose a path f from x to y, $f : [0,1] \longrightarrow X$. Since $[0,1]$ is connected, $f([0,1])$ is connected, by 3.6 on page 105. So x and y (which are in the image of f) lie in the same component of x. Since x and y were arbitrary, X is connected. □

4.10 Result

Let X have the indiscrete topology.

Claim: X is path connected.

Proof. Let $x, y \in X$. Define $f : [0,1] \longrightarrow X, f(0) = x, f(t) = y \ \forall \ t \in (0,1]$. By 2.10 on page 83, f is continuous. f is a path from x to y. So X is path connnected. □

Part VI

Separation axioms

Chapter 1

T_1

1.1 Definition

Let X be a space. We say X is T_1 (read 'tee - one') if $\{x\}$ is closed in X for every $x \in X$.

1.2 Example

Let $\mathbb{R}$ have the standard topology.

Claim: $\mathbb{R}$ is T_1.

Proof. 4.6 on page 61 □

1.3 Example

Let X be a space, with the cofinite topology.

Claim: X is T_1.

Proof. 4.4 on page 60 □

1.4 Result

Let X be a space.

Claim: X is $T_1 \iff \forall$ distinct $a, b \in X \,\exists\, U_a, U_b$ (neighborhoods of a and b respectively) such that $a \notin U_b$ and $b \notin U_a$.

Proof. $\implies$ Suppose X is T_1. Let $a, b \in X, a \neq b$. $\{a\}, \{b\}$ are closed, since X is T_1. Let $U_a = X - \{b\}$ and $U_b = X - \{a\}$. $a \in U_a$ and $b \in U_b$. U_a and U_b are open. $b \notin U_a, a \notin U_b$.

$\impliedby$ Suppose $\forall$ distinct $a, b \in X \,\exists\, U_a, U_b$ (neighborhoods of a and b respecctively) such that $a \notin U_b$ and $b \notin U_a$. Let $x \in X$. We want to show $\{x\}$ is closed. So we want to show that $X - \{x\}$ is open. Let $y \in X - \{x\}$. If we can show y is an interior point of $X - \{x\}$, that would be nice. Then every point of $X - \{x\}$ is an interior point of $X - \{x\}$, and thus $X - \{x\}$ would be open by 5.5 on page 66. And $\{x\}$ would be closed. And we would be done.

Lemma: y is an interior point of $X - \{x\}$

Proof of Lemma. Since $x \neq y$, we have U_x and U_y neighborhoods of x and y with $x \notin U_y$, $y \notin U_x$. Since $x \notin U_y$, $U_y \subset X - \{x\}$. So we have a neighborhood of y that is contained in $X - \{x\}$. So y is an interior point of $X - \{x\}$. □

So we are done. □

1.5 Result

Let T be the lower limit topology on $\mathbb{R}$.

Claim: $(\mathbb{R}, T)$ is T_1.

Part VI: Separation axioms

Proof. Let $x \in \mathbb{R}$. We want to show $\{x\}$ is closed in $\mathbb{R}$. We will show $\mathbb{R} - \{x\}$ is open. $\mathbb{R} - \{x\} = (-\infty, x) \bigcup (x, \infty)$. By 2.6 on page 121 and 2.8 on page 121 $\mathbb{R} - \{x\}$ is open. So $\{x\}$ is closed, and $(\mathbb{R}, T)$ is T_1. $\square$

1.6 Result

Let X be T_1. Let $A \subset X$, A equipped with the subspace topology.

Claim: A is T_1.

Proof. Let $a \in A$. $\{a\}$ is closed in X. So $X - \{a\}$ is open in X. So $A \bigcap (X - \{a\})$ is open in A. But $A \bigcap (X - \{a\}) = A - \{a\}$ (Why?). So $A - \{a\}$ is open in A. So $\{a\}$ is closed in A. So A is T_1. $\square$

Chapter 2

T_2 - Hausdorff Spaces

2.1 Definition

Let X be a space. We say X is T_2 (or **Hausdorff**) when the following is true: $x, y \in X$ and $x \neq y \implies \exists$ disjoint open sets U and V such that $x \in U$ and $y \in V$.

2.2 Result

Let X be T_2.

Claim: X is T_1.

Proof. Case 1: $X = \{x\}$.
Then $\{x\} = X$ is closed. So X is T_1.

Case 2: X has more than one element.
Let $x \in X$. We want to show that $\{x\}$ is closed. So we want to show $X - \{x\}$ is open. Let $y \in X - \{x\}$. Since X is T_2, we know $\exists U, V$ open, disjoint such that $x \in U$ and $y \in V$. V is a neighborhood of y, and $V \subset X - \{x\}$. So y is an interior point of $X - \{x\}$.

So for an arbitrary y, we have shown $y \in X - \{x\}$ is an interior point of $X - \{x\}$. So, by 5.5 on page 66, $X - \{x\}$ is open. So $\{x\}$ is closed.

So X is T_1.

□

2.3 Result

Let $\mathbb{R}$ have the standard topology.

Claim: $\mathbb{R}$ is T_2.

Proof. Let $a, b \in \mathbb{R}$, $a \neq b$. Without loss of generality, assume that $a < b$. Let $U = (a - 1, \frac{a+b}{2})$, $V = (\frac{a+b}{2}, b + 1)$. U and V are open. U and V are disjoint. $a \in U$, $b \in B$. So $\mathbb{R}$ is T_2.

□

2.4 Result

Let X be a Hausdorff space, $Y \subset X$.

Claim: Y is Hausdorff.

Proof. Let $a, b \in Y$, $a \neq b$. Since $a, b \in X$, $\exists$ disjoint neighborhoods (open in X) of a and b, U and V. $Y \bigcap U$, $Y \bigcap V$ are disjoint, non-empty, and neighborhoods of a and b (open in Y). So Y is Hausdorff.

□

2.5 Result

Let X, Y be spaces, Y Hausdorff, $f : X \longrightarrow Y$ a continuous, 1-1 function.

Part VI: Separation axioms 139

Claim: X is Hausdorff.

Proof. Let $x, y \in X$, $x \neq y$. Since f is 1-1, $f(x) \neq f(y)$. So we have disjoint open sets U, V such that $f(x) \in U, f(y) \in V$. Since f is continuous, $f^{-1}(U)$ and $f^{-1}(V)$ are open. Since $f(x) \in U$ and $f(y) \in V$, $x \in f^{-1}(U)$ and $y \in f^{-1}(V)$.

Lemma: $f^{-1}(U) \bigcap f^{-1}(V) = \phi$.

Proof of Lemma: Suppose $f^{-1}(U)$ and $f^{-1}(V)$ are not disjoint. Then $\exists\, a \in f^{-1}(U) \bigcap f^{-1}(V)$. So $a \in f^{-1}(U)$ and $a \in f^{-1}(V)$. So $f(a) = b \in U$ and $f(a) = c \in V$. But f is well-defined, so $b = c$. That's bad, since it implies $b \in U \bigcap V = \phi$. So our assumption that U and V are not disjoint has led to a contradiction. So U and V are disjoint.

□

□

2.6 Result

Claim: Hausdorff is a topological property.

Proof. 2.5 (Why?) □

2.7 Result

Let (X, T) be a Hausdorff space. Let T' be a topology on X with $T \subset T'$.

Claim: (X, T') is Hausdorff.

Proof. Let $a, b \in X$, $a \neq b$. Then since (X, T) is Hausdorff, we have disjoint $U, V \in T$ such that $a \in U$, $b \in V$. And since $T \subset T'$, $U, V \in T'$. So (X, T') is Hausdorff. □

2.8 Result

Let T be the lower limit topology on $\mathbb{R}$.

Claim: $(\mathbb{R}, T)$ is Hausdorff.

Proof. Let $a, b \in \mathbb{R}$, $a \neq b$. Without loss of generality, assume $a < b$. Let $x = \frac{a+b}{2}$. Then $a \in [a, x)$ and $b \in [x, b+1)$. $[a, x)$ and $[x, b+1)$ are disjoint open sets. So $\mathbb{R}$, with the lower limit topology, is Hausdorff. □

2.9 Result

Let T be the K-topology on $\mathbb{R}$.

Claim: $(\mathbb{R}, T)$ is Hausdorff.

Proof. This is trivial by 2.11 on page 122, 2.3 on page 138, and 2.7 on the previous page. Why? □

2.10 Result

Let $f, g : X \longrightarrow Y$ be continuous. Let Y be T_2. Let $A = \{x \in X;\ f(x) = g(x)\}$.

Claim: A is closed in X.

Proof. We show that $X - A$ is open in X. $X - A = \{x \in X;\ f(x) \neq g(x)\}$. Let $a \in X - A$. Then $f(a) \neq g(a)$ and $f(a), g(a) \in Y$. Since Y is T_2, we have open, disjoint U, V such that $f(a) \in U$, $g(a) \in V$. So $a \in f^{-1}(U)$ which is open by continuity of f. And $a \in g^{-1}(V)$ which is open by continuity of g. Let $W = f^{-1}(U) \bigcap g^{-1}(V)$. W is open. And $a \in W$. It would be nice if $W \subset X - A$. If that were true, then we will have found a neighborhood W of a, contained

Part VI: Separation axioms 141

in $X - A$. That would make a an interior point of $X - A$. Since a was arbitrary, that would make $X - A$ open. So A would be closed.

Lemma: $W \subset X - A$

Proof of Lemma: We show $W \cap A = \phi$. Suppose $x \in W \cap A$. Then since $x \in W$, we have $f(x) \in U$, $g(x) \in V$. But, since $x \in A$, we have $f(x) = g(x)$. This implies $U \cap V \neq \phi$. But U and V are disjoint. So our assumption that $W \cap A \neq \phi$ is false. So $W \cap A = \phi$.

□

□

2.11 Result

Let X be T_2. Let Y be a retract of X.

Claim: Y is closed in X.

Proof. We show $X - Y$ is open in X. Let $x \in X - Y$. We want to show x is an interior point of $X - Y$.
Since Y is a retract of X, we have continuous $r : X \longrightarrow Y$ with $r(y) = y \, \forall \, y \in Y$.
$r(x) \in Y$. Since X is T_2, $\exists \, U, V^{open} \subset X$, with $U \cap V = \phi$ and $x \in U$, $r(x) \in V$. Since $V^{open} \subset X$, $(V \cap Y)^{open} \subset Y$. Since r is continuous, $r^{-1}(V \cap Y)$ is open in X.
Since $r(x) \in V$ and $r(x) \in Y$, $x \in r^{-1}(V)$ and $x \in r^{-1}(Y)$. So $x \in r^{-1}(V \cap Y)$.

So we have $x \in U$ and $x \in r^{-1}(V \cap Y)$.
So $x \in U \cap (r^{-1}(V \cap Y))$, which is open. It would be nice if $U \cap (r^{-1}(V \cap Y)$ and Y were disjoint. In that case we would have shown that x is an interior point of $X - Y$ and we would

be done.

Lemma: $(U \cap (r^{-1}(V \cap Y))) \cap Y = \phi$

Proof of Lemma: Suppose not. Then $x \in U$, $x \in r^{-1}(V \cap Y)$ and $x \in Y$. So $r(x) \in V$ and $r(x) \in Y$ and $x \in Y$. Since $x \in Y$, $r(x) = x$. But $r(x) \in V$. So $x \in V$. So we have $x \in U \cap V = \phi$, which is bad. So our assumption that $U \cap (r^{-1}(V \cap Y))$ and Y are not disjoint has led to a contradiciton. So $U \cap (r^{-1}(V \cap Y))$ and Y are disjoint. □

By the above comments we are done. □

Chapter 3

T_3 - Regular Spaces

3.1 Definition

Let X be a space. We say X is T_3 (or **regular**) when X is T_1 and the following is true:
$x \in X$, $C^{closed} \subset X$, and $x \notin C \implies \exists$ disjoint open sets U and V such that $x \in U$, and $C \subset V$.

3.2 Result

Let X be T_3.

Claim: X is T_2.

Proof. Let $x, y \in X$. Since X is T_1, we know $\{x\}$ is closed. Since X is regular, $\exists U, V$ open, disjoint such that $\{x\} \subset U$ and $y \in V$. And since $\{x\} \subset U$, $x \in U$. So we have shown X is T_2. $\square$

3.3 Example

Let T be the standard topology on $\mathbb{R}$.

Claim: $(\mathbb{R}, T)$ is T_3.

Proof. Let $a \in \mathbb{R}$, $C^{closed} \subset \mathbb{R}$, $a \notin C$. By 5.6 on page 67, a is not a limit point of C. So $\exists\, W$ a neighborhood of a, $W \subset \mathbb{R} - C$. We have a basis for T, all open intervals in $\mathbb{R}$. So W is a union of open intervals in $\mathbb{R}$. So, we can find an open interval in $\mathbb{R}$, (x, y) so that $a \in (x, y) \subset W \subset \mathbb{R} - C$. Also, we have $a \in (\frac{x+a}{2}, \frac{a+y}{2})$, and $C \subset (-\infty, \frac{x+a}{2}) \bigcup (\frac{a+y}{2}, \infty)$. Let $U = (\frac{x+a}{2}, \frac{a+y}{2})$, $V = (-\infty, \frac{x+a}{2}) \bigcup (\frac{a+y}{2}, \infty)$. U and V are open, disjoint sets. And $a \in U$, $C \subset V$. $\square$

3.4 Result

Claim: T_3 is a topological property.

Proof. Let $f : X \longrightarrow Y$ be a homeomorphism. Let X be T_3. Let $A^{closed} \subset Y$, $y \in Y - A$. Then $f^{-1}(A)^{closed} \subset X$. And $f^{-1}(y) \in X - f^{-1}(A)$ (Why?). Since X is regular, we have disjoint open sets $U, V \subset X$ such that $f^{-1}(y) \in U$ and $f^{-1}(A) \subset V$. $f(U)$ and $f(V)$ are open and disjoint (Why?). And $y \in f(U)$, $A \subset f(V)$. So Y is regular. So T_3 is a topological property. $\square$

3.5 Definition

Let X be a space. We say X is T_4 (or **normal**) when the following is true:

A and B closed subsets of $X \implies \exists$ disjoint open sets U and V such that $A \subset U$, and $B \subset V$.

Chapter 4

Compact Spaces

4.1 Definition

Let X be a space, and let $\{U_\alpha\}$ be a collection of open sets in X. Then we say $\{U_\alpha\}$ is an **open covering** of X if $\forall\ x \in X$, $x \in U_b$ for some $U_b \in \{U_\alpha\}$.

4.2 Example

Let's look at $\mathbb{R}$ with the standard topology.
Let $U_n = (-n,\ n)$. Then $\{U_n\}_{n \in \mathbb{N}}$ is an open covering of $\mathbb{R}$.

4.3 Definition

Let X be a space and let $\{U_\alpha\}$ be an open covering of X. A **subcover** of $\{U_\alpha\}$ is a collection of some of the elements of $\{U_\alpha\}$ that forms an open cover of X.

4.4 Definition

Let X be a space and let $\{U_\alpha\}$ be an open covering of X. A **finite subcover** of $\{U_\alpha\}$ is a subcover of $\{U_\alpha\}$ that has a finite number of elements.

4.5 Definition

Let X be a space. X is said to be **compact** when the following is true: For any open covering $\{U_\alpha\}$ of X, $\exists \{U_1, \ldots, U_n\}$ a finite subcover of $\{U_\alpha\}$.

4.6 Problem

Let X be a finite topological space. Show X is compact.

4.7 Problem

Show $[0, 1] \subset \mathbb{R}$ is compact.

4.8 Problem

Show $[0, 1] \subset \mathbb{R}$ is not homeomorphic to $(0, 1) \subset \mathbb{R}$.

4.9 Result

Let $A^{closed} \subset X^{compact}$. Show A is compact.

Proof. Let J be a covering of A by sets open in X. Then $J \bigcup \{X - A\}$ is an open covering of X. So, some finite subcollection of $J \bigcup \{X - A\}$ covers X. This finite subcover will also cover A. So A is compact. $\square$

4.10 Problem

Let $A^{compact} \subset X^{T_2}$. Show A is closed.

4.11 Problem

Equip $\mathbb{R}$ with the standard topology. Let $a, b \in \mathbb{R}, a < b$. Show none of the following are compact in $\mathbb{R}$:
$(a, b), [a, b), (a, b]$

4.12 Problem

Consider $[0, 1], (0, 1), [0, 1), (0, 1]$ as subspaces of $\mathbb{R}$ with the standard topology.
1) Show $[0, 1]$ is not homeomorphic to $(0, 1)$.
2) Show $[0, 1]$ is not homeomorphic to $[0, 1)$.
3) Show $[0, 1]$ is not homeomorphic to $(0, 1]$.

4.13 Problem

Show $\mathbb{R}$ is not compact.

Chapter 5

Metric Spaces

5.1 Definition

Let $f : X \times X \longrightarrow \mathbb{R}$. We say f is a **metric** on X when the following are satisfied.

1) $f(x,y) \geq 0 \,\forall\, x,y \in X$.

2) $f(x,y) = 0 \iff x = y$.

3) $f(x,y) = f(y,x) \,\forall\, x,y \in X$.

4) $f(x,y) + f(y,z) \geq f(x,z) \,\forall\, x,y,z \in X$.

5.2 Definition

Let X be a set with a metric f. Let $x \in X$. We define a **ball** about x of radius a, $B(x,a) = \{y \in X;\ f(x,y) < a\}$, where $a \in (0, \infty) \subset \mathbb{R}$.

5.3 Problem

Let X be a set with a metric d.
Let $B = \{B(x,a); x \in X, a > 0\}$. Show B is a basis for a topology T on X.

5.4 Definition

Let X be a set with a metric d.
Let $B = \{B(x,a); x \in X, a > 0\}$. Then B is a basis for a topology on X called the **metric topology** on $\mathbb{R}$. When X is equipped with this topology, we say X is a **metric space**.

5.5 Result

Let X be a set.
Define $d : X \times X \longrightarrow \mathbb{R}$, $d(x,y) = \begin{cases} 0 & \text{for } x = y \\ 1 & \text{for } x \neq y \end{cases}$

Claim: d is a metric on X.

Proof:
1) $d(x,y) \geq 0 \;\forall\, x, y \in X$

2) $d(x,y) = 0 \iff x = y$

3) $d(x,y) = d(y,x) \;\forall\, x, y \in X$

4) Let $x, y, z \in X$.
$d(x,z) = 0$ or $d(x,z) = 1$.

Case 1: $d(x,z) = 0$
Then $d(x,z) \leq d(x,y) + d(y,z)$.

Part VI: Separation axioms   151

Case 2: $d(x,z) = 1$
Then $x \neq z$. Suppose $d(x,z) \not\leq d(x,y) + d(y,z)$.
Then $d(x,y) + d(y,z) = 0$. So $d(x,y) = 0$, and $d(y,z) = 0$.
So $x = y$, and $y = z$. So $x = z$. But $x \neq z$.
So $d(x,z) \leq d(x,y) + d(y,z)$. So d is indeed a metric on X.

5.6 Problem

Let $d : \mathbb{R} \times \mathbb{R} \longrightarrow \mathbb{R} = |y - x|$. Show $\mathbb{R}$ is a metric space, with metric d.

5.7 Problem

Let T be the standard topology on $\mathbb{R}$. Let T' be the topology induced on $\mathbb{R}$ by the metric d in 5.6. Show $T = T'$.

5.8 Result

Let X have metric d.

Claim: X is T_1.

Proof. Let $x \in X$. We want to show $\{x\}$ is closed. We will show $X - \{x\}$ is open. Let $y \in X - \{x\}$. Since $y \neq x$, we have $d(x,y) = t > 0$. $B(y, \frac{t}{2})$ is open. It would be nice if we had $B(y, \frac{t}{2}) \bigcap \{x\} = \phi$. Then we would have a neighborhood of y contained in $Y - \{x\}$. So y would be an interior point of $X - \{x\}$. And $X - \{x\}$ would be open. And $\{x\}$ would be closed.

Lemma: $B(y, \frac{t}{2}) \bigcap \{x\} = \phi$

Proof of Lemma. Suppose not. Then $x \in B(y, \frac{t}{2})$. So $d(x,y) < \frac{t}{2}$. But $d(x,y) = t > \frac{t}{2}$.

□

By comments above, we are done. □

5.9 Result

Let X have metric d.

Claim: X is T_4.

Proof. Let A, B be disjoint closed subsets of X. $X - B$ is open. Since $A \subset X - B$, for each $a \in A \; \exists \; \epsilon_a > 0$ such that $B(a, \epsilon_a) \subset X - B$. Since $X - A$ is open and $B \subset X - A$, for each $b \in B \; \exists \; \epsilon_b > 0$ such that $B(b, \epsilon_b) \subset X - A$. Let $U = \bigcup_{a \in A} B(a, \frac{\epsilon_a}{2})$ and let $V = \bigcup_{b \in B} B(b, \frac{\epsilon_b}{2})$. U and V are open, since each $B(a, \epsilon)$ is open.

It would be nice if $A \subset U$, $B \subset V$, and $U \bigcap V = \phi$. Then we would be done.

Lemma 1: $A \subset U$

Proof of Lemma 1. Let $t \in A$. Then $t \in B(t, \frac{\epsilon_t}{2}) \subset \bigcup_{a \in A} B(a, \frac{\epsilon_a}{2}) = U$. So $t \in U$.

□

Lemma 2: $B \subset V$

Proof of Lemma 2. Similar to Lemma 1.

□

Lemma 3: $U \bigcap V = \phi$.

Part VI: Separation axioms   153

Proof of Lemma 3. Suppose not. Then $\exists\, x \in U \cap V$. So $x \in (\bigcup_{a \in A} B(a, \frac{\epsilon_a}{2})) \cap (\bigcup_{b \in B} B(b, \frac{\epsilon_b}{2}))$. So for some fixed $a \in A$ and $b \in B$, $x \in B(a, \frac{\epsilon_a}{2}) \cap B(b, \frac{\epsilon_b}{2})$.
$d(a, b) \leq d(a, x) + d(x, b)$.
But $d(a, x) < \frac{\epsilon_a}{2}$ and $d(x, b) < \frac{\epsilon_b}{2}$.
So $d(a, b) < \frac{\epsilon_a}{2} + \frac{\epsilon_b}{2} = \frac{(\epsilon_a + \epsilon_b)}{2}$.

Case 1: $\epsilon_a \leq \epsilon_b$
Then $d(a, b) < \frac{(\epsilon_b + \epsilon_b)}{2} = \epsilon_b$. So $a \in B(b, \epsilon_b)$.
But $B(b, \epsilon_b) \subset X - A$. contradiction.

Case 2: $\epsilon_b > \epsilon_a$
Then $d(a, b) < \frac{(\epsilon_a + \epsilon_a)}{2} = \epsilon_a$. So $b \in B(a, \epsilon_a)$.
But $B(a, \epsilon_a) \subset X - B$. contradiction.

So $U \cap V = \phi$ □

As discussed above, by the Lemmas we are done.   □

5.10 Problem

Let X be a metric space with metric d. Let $A \subset X$, equip A with the subspace topology. Show A is also a metric space, with metric the restriction of d to A.

5.11 Result

Let X be a metric space with metric g and Y a metric space with metric h.
Let $f : X \longrightarrow Y$.

Claim: f is continuous $\iff$ for any $x \in X$ and $\epsilon > 0\, \exists\, \delta > 0$ such that $g(x, a) < \delta \implies h(f(x), f(a)) < \epsilon\, \forall\, a \in X$.

Proof. $\implies$ Suppose f is continuous. Let $\epsilon > 0$ and $x \in X$. $B(f(x), \epsilon)$ is open in Y. Since f is continuous, $f^{-1}(B(f(x), \epsilon))$ is open in X. But $x \in f^{-1}(B(f(x), \epsilon))$. So there is a basis element $B(x, \delta)$ such that $x \in B(x, \delta) \subset f^{-1}(B(f(x), \epsilon))$.

Suppose $a \in X$ and $g(x, a) < \delta$. Then $a \in B(x, \delta) \subset f^{-1}(B(f(x), \epsilon))$. So $f(a) \in B(f(x), \epsilon)$. So $h(f(x), f(a)) < \epsilon$.

$\impliedby$ Suppose for any $x \in X$ and $\epsilon > 0$ $\exists \delta > 0$ such that $g(x, a) < \delta \implies h(f(x), f(a)) < \epsilon \ \forall \, a \in X$. We want to show f is continuous. Let $U^{open} \subset Y$. We want to show $f^{-1}(U)^{open} \subset X$. Let $x \in f^{-1}(U)$. It would be nice if x were an interior point of $f^{-1}(U)$. Then $f^{-1}(U)$ would be open and f would be continuous and we would be done.

Lemma: x is an interior point of $f^{-1}(U)$.

Proof of Lemma: $f(x) \in U^{open}$ (since $x \in f^{-1}(U)$). So $\exists \epsilon > 0$ such that $B(f(x), \epsilon) \subset U$. So $\exists \delta > 0$ such that $g(x, a) < \delta \implies h(f(x), f(a)) < \epsilon \ \forall \, a \in X$. That is, $f(B(x, \delta)) \subset B(f(x), \epsilon)$. So $B(x, \delta) \subset f^{-1}(B(f(x), \epsilon)) \subset f^{-1}(U)$. $B(x, \delta)$ is open and $x \in B(x, \delta)$. So x is an interior point of $f^{-1}(U)$. $\square$

We are done. $\square$

5.12 Remark

We can say more true things about the words we have defined in the book. We can also define more words, and say true things about them. That characterizes many math problems not given in this book.

Index

1-to-1, 28

ball, 149
basis, 73
bijective, 30

closed, 59
codomain, 27
cofinite, 53
cofinite topology, 54
compact, 146
complement, 22
component, 108
composition, 32
connected, 103
contained, 18
continuous, 79

discrete topology, 46
disjoint, 103
domain, 27

element, 17
empty set, 18
equality of functions, 32
equality of sets, 21

finite, 23
finite subcover, 146

fixed point, 99
Fixed Point Property, 99
function, 27

Hausdorff, 137
homeomorphic, 87
homeomorphism, 87

identity, 32
image, 36
inclusion, 85
indiscrete topology, 45
infinite, 24
injective, 28
interior point, 65
intersection, 20
inverse, 33
inverse image, 36

K-topology, 121

limit point, 66
lower limit topology, 119

metric, 149
metric space, 150
metric topology, 150

neighborhood, 46

155

normal, 144

onto, 29
open, 43
open covering, 145
open interval, 47

path, 127
path connected, 127
point, 47

regular, 143
restriction, 85
retract, 95
retraction, 95

separation, 103
set, 17
Sierpinski space, 123
space, 43
standard topology on $\mathbb{R}$, 49
subcover, 145
subset, 18
subspace topology, 114
surjective, 29

T_1, 133
T_2, 137
T_3, 143
T_4, 144
topological property, 91
topological space, 43
topology, 43

union, 20

well-defined function, 27

Bibliography

[1] M.A. Armstrong, *Basic topology*, Springer, 1997.

[2] V.L.Gutenmacher A.T.Fomenko, D.B.Fukes, *Homotopic topology*, H Stillman Pub, 1986.

[3] Ronald Brown, *Elements of modern topology*, McGraw-Hill Education, 1968.

[4] Gerard Buskes and Arnoud van Rooij, *Topological spaces: From distance to neighborhood*, Springer, 1997.

[5] Stephan C. Carlson, *Topology of surfaces, knots, and manifolds*, Wiley, 2001.

[6] J. A. Thorpe I. M. Singer, *Lecture notes on elementary topology and geometry*, Springer, 1976.

[7] Irving Kaplansky, *Set theory and metric spaces*, American Mathematical Society, 2001.

[8] J. Arthur Seebach Lynn Arthur Steen, *Counterexamples in topology*, Dover Publications, 1995.

[9] James Munkres, *Topology*, Prentice Hall, 2000.

[10] Charles Pitts, *Introduction to metric spaces*, Oliver and Boyd, 1972.

[11] Dennis Roseman, *Elementary topology*, Prentice Hall, 1999.

Contact Information

Thank you for your purchase of this Bobo Strategy publication.

For more information about the products and services we offer, please contact us by mail at:

Bobo Strategy
2506 N Clark # 287
Chicago, IL 60614

Or visit our website at http://www.bobostrategy.com

Or send us an email at info@bobostrategy.com

We hope you have found this product both enjoyable and useful.

www.ingramcontent.com/pod-product-compliance
Lightning Source LLC
Chambersburg PA
CBHW032259150426
43195CB00008BA/509